Prentice Hall Advanced Reference Series

Physical and Life Sciences

PRENTICE HALL
Biophysics and Bioengineering Series
Abraham Noordergraaf, Series Editor

AGNEW AND MCCREERY, EDS. *Neural Prostheses: Fundamental Studies*
ALPEN *Radiation Biophysics*
DAWSON *Engineering Design of the Cardiovascular System of Mammals*
GANDHI, ED. *Biological Effects and Medical Applications of Electromagnetic Energy*
LLEBOT AND JOU *Introduction to the Thermodynamics of Biological Processes*
RIDEOUT *Mathematical and Computer Modeling of Physiological Systems*

FORTHCOMING BOOKS IN THIS SERIES *(tentative titles)*

COLEMAN *Integrative Human Physiology: A Quantitative View of Homeostasis*
FOX *Fundamentals of Medical Imaging*
GRODZINSKY *Fields, Forces, and Flows in Biological Tissues and Membranes*
HUANG *Principles of Biomedical Image Processing*
MAYROVITZ *Analysis of Microcirculation*
SCHERER *Respiratory Fluid Mechanics*
VAIDHYANATHAN *Regulation and Control in Biological Systems*
WAAG *Theory and Measurement of Ultrasound Scattering in Biological Media*

ENGINEERING DESIGN OF THE CARDIOVASCULAR SYSTEM OF MAMMALS

Thomas H. Dawson
Professor of Engineering
United States Naval Academy
Annapolis, Maryland

Prentice Hall
Englewood Cliffs, New Jersey 07632

Dawson, Thomas H.
   Engineering design of the cardiovascular system of mammals /
Thomas H. Dawson.
      p.   cm. -- (Prentice Hall advanced reference series. Physical
and life sciences) (Prentice Hall biophysics and bioengineering
series)
   Includes index.
   ISBN 0-13-275694-3
   1. Cardiovascular system. 2. Bioengineering. 3. Hemodynamics.
4. Blood--Circulation.   I. Title.   II. Series.   III. Series:
Prentice Hall biophysics and bioengineering series.
QP105.D38 1991
599'.011--dc20   90-24190
                                                              CIP

*COPM
QP
105
. D38
1991
c. 1*

Editorial/production supervision: Rick DeLorenzo
Interior design: Karen Bernhaut
Cover design: Lundgren Graphics
Manufacturing buyer: Patrice Fraccio
Acquisitions editor: Ken Tennity

Prentice Hall Advanced Reference Series
Prentice Hall Biophysics and Bioengineering Series

© 1991 by Prentice-Hall, Inc.
A Division of Simon & Schuster
Englewood Cliffs, New Jersey 07632

Printed in the United States of America
10 9 8 7 6 5 4 3 2 1

ISBN 0-13-275694-3

Prentice-Hall International (UK) Limited, *London*
Prentice-Hall of Australia Pty. Limited, *Sydney*
Prentice-Hall Canada Inc., *Toronto*
Prentice-Hall Hispanoamericana, S.A., *Mexico*
Prentice-Hall of India Private Limited, *New Delhi*
Prentice-Hall of Japan, Inc., *Tokyo*
Simon & Schuster Asia Pte. Ltd., *Singapore*
Editora Prentice-Hall do Brasil, Ltda., *Rio de Janeiro*

To Lois, Tamalyn, and Tephanie

Contents

PREFACE   xi

ACKNOWLEDGMENTS   xiii

UNITS AND ABBREVIATIONS   xv

1. INTRODUCTION   1

   1.1.   Scaling Laws   2
   1.2.   Implications to Design   7
   1.3.   Heat Production Theory   9
   1.4.   Dimensional Analysis   13
   1.5.   Summary   17

2. GENERAL FEATURES OF THE CARDIOVASCULAR SYSTEM   19

   2.1.   Organization   19
   2.2.   The Working Fluid—The Blood   21
   2.3.   The Pumps—The Heart   34
   2.4.   The Piping Network—The Vascular Beds   48

3. DESIGN THEORY AND SIMILARITY REQUIREMENTS   61

   3.1.   Engineering Design Model   62
   3.2.   Similarity Requirements   66
   3.3.   Additional Design Assumptions   69
   3.4.   Scaling Laws   71

4. APPLICATIONS OF DESIGN THEORY   **75**

    4.1.   Geometric Scaling Relations   75
    4.2.   Blood Motions and Pressures   81
    4.3.   Scaling Laws During Growth and Aging   88
    4.4.   Oxygen Partial Pressures in Blood   93
    4.5.   Capillary Density in Muscle Tissue   98

5. SPECIAL CONSIDERATION OF INDIVIDUAL ORGANS   **103**

    5.1.   Oxygen Uptake of Organs   103
    5.2.   Blood Flow to Organs   110
    5.3.   Vascular Design of Kidneys   114
    5.4.   Lung Design   121
    5.5.   Mechanical Design of the Heart   130

6. CARDIAC RESPONSE CHARACTERISTICS   **135**

    6.1.   Simplified Governing Equation   135
    6.2.   Predictions for the Resting Human   137
    6.3.   Cardiac Response to Exercise   140

7. REVIEW AND REFLECTION   **149**

    7.1.   Review and Summary   149
    7.2.   Design Methodology   154
    7.3.   Implications and Future Work   158

APPENDIX A.   **LOGARITHMS AND SCALING EQUATIONS**   **161**

APPENDIX B.   **SCALING LAWS FOR VOLUMES AND AREAS**   **165**

APPENDIX C.   **DIMENSIONAL ANALYSIS AND PHYSICAL SCALING LAWS**   **167**

APPENDIX D.   **SOME PHYSICAL CONCEPTS**   **171**

INDEX   **177**

Foreword

Originally in the form of a collection of clay tablets, then as a papyrus roll, and eventually as bound sheets of paper, the book has served to record events and ideas uppermost in the contemporary human mind for the last 45 to 50 centuries.

In one familiar classic role it continues as a teaching aid in the instruction of the next generation: the textbook. The noble textbook may have several precursors.

In the last few centuries, as new research areas are discovered or older ones become accessible owing to the creation of new methods, the professional journals have become the initial repositories of reports that treat the various aspects in excruciating detail. As the number of publications multiplies, it tends to become a challenge to find and especially to digest such a collection of papers owing to real or apparent contradictions in measurement results or promulgated interpretations. A well-organized and critical collection in book form tends to crystallize the issues and purify the interpretation, thereby facilitating further progress in that research area.

Once the proper methodology or the governing principle has been identified, a book can serve to set forth the solution, thus reducing the volume of current literature by a significant factor.

Eventually, a book tracing the generation and growth of ideas in a broader field can serve to expose guiding philosophical convictions over a more extended period of research, often resulting in striking insights.

The Biophysics and Bioengineering Series is intended to serve these several purposes in the broad field defined in its name by providing monographs prepared by expert investigators. Originally published by Academic Press, the series is continued by Prentice Hall.

Abraham Noordergraaf
Series Editor

Preface

The cardiovascular system is that complex apparatus of the body responsible for the circulation of life-sustaining blood through all of its many parts. In addition to the blood, the system consists of the heart, the arteries and veins, and the various capillary networks spread throughout the tissues. The heart provides the pumping action needed in the circulation; the arteries and veins provide the closed vessels needed for directing the circulation; and the capillaries provide the transfer devices needed for exchange of products with surrounding tissues.

The system is normally thought of in connection with the circulation of blood in humans. Interestingly enough, however, there exists a striking similarity in the behavior of the system for all mammals. For example, the average heart rates of the mouse, man, and the elephant differ significantly from one another, yet all can be computed from a single algebraic equation involving only the animal weight. Similar so-called scaling relations are known, or can be developed, for virtually all the measured physiological variables of the system.

The existence of such relations implies that the design of the cardiovascular system of all mammals is governed by the same basic design theory. By design, we mean here simply the makeup of the system as we find it from experimental observations. Similarly, by design theory, we mean the fundamental principles governing the design and operation of the system. Although many scaling relations have been developed from experimental measurements, no past effort has yet resulted in identifying this design theory and the associated design equations that make these relations possible.

This brings us to the subject of the present book, namely, a consideration of the design of the cardiovascular system from an engineering perspective. By applying relatively simple engineering concepts and arguments to the study of the system, it is shown possible here to develop a design theory that does, indeed, yield the known scaling relations of the kind described above. More importantly, the theory also reveals additional scaling relations not previously known and provides new insight into the design and fundamental operation of the system.

The work described here is original in that its essential parts have not previously been published. It is exciting because, in understanding the basis for the various scaling relations among mammals, much new understanding is gained about our own system's operation. And it is humbling in the sense that the human system is found to be neither more nor less advanced than that of the mouse, the pig, or the elephant.

In developing the subject, I have kept the mathematics to a simple level in order to make the physical description as clear as possible. The reader with an interest in physiology and a basic knowledge of algebra should have no difficulty following the presentation. The reader with a background in science and engineering and an interest in living systems will likewise find no trouble in following the presentation and appreciating the engineering parallels.

ACKNOWLEDGMENTS

I am grateful to a number of people for their assistance with this book. For their efforts and encouragement, I particularly wish to thank Mr. Kenneth Tennity, Senior Managing Editor for Advanced Sciences at Prentice-Hall and Dr. Abraham Noordergraaf, Professor at the University of Pennsylvania and Technical Editor of the Prentice-Hall Biophysics and Bioengineering Series.

I also wish to thank the several anonymous reviewers who provided valuable comments and encouragement both initially and in the final stages of preparation of the work.

The excellent contributions to the book by Mrs. Jackie Martin, Mr. Rick DeLorenzo, and the staff of the Editorial-Production Department at Prentice-Hall are likewise appreciated.

At the Naval Academy, I wish to thank my friend and colleague, Professor Thomas W. Butler, for his interest in the work and for reviewing and commenting on various portions of it.

Thomas H. Dawson

Units and Abbreviations

QUANTITY	MEASURE	ABBREVIATION
mass or weight*	kilogram	kg
	gram	g
length	meter	m
	centimeter	cm
	millimeter	mm
time	minute	min.
	second	sec
volume	liter	l
	milliliter	ml
heat	kilocalorie	kcal
force*	dyne	
	gram	g
pressure	millimeters of mercury	mm Hg
potential	volts	

*Strictly, kilogram and gram are units of mass. When used as a weight or force measure, they denote the weight or downward force experienced by that amount of mass.

1

Introduction

The cardiovascular system of mammals consists broadly of the blood, the heart, and the various blood vessels of the body. Its function is to provide blood flow throughout the body so that oxygen and nutrients can be supplied to the cells and waste products removed. The system accordingly involves mainly a mechanical plant so organized as to circulate the blood through the various parts of the body and allow exchange of products with its surroundings.

From an engineering perspective, the essential components of the system may be categorized as (1) a working fluid, the blood; (2) a dual pulsatile pump, the heart; (3) a closed piping system, the arteries and veins; and (4) numerous diffusion devices, the capillaries of the various parts of the body. With this system, oxygen-enriched blood is pumped from the left side of the heart through a series of branching arteries to the capillaries of the tissues where exchange of oxygen, carbon dioxide, and other products take place by diffusion, filtration, and absorption processes. The blood is then returned to the right side of the heart by the veins, where it is pumped to the lungs for gas exchange and then returned to the left side of the heart for recirculation.

It is noteworthy that both the organization and operation of the cardiovascular system are similar for virtually all mammals. Insofar as organization is concerned, all mammals have, for example, a four-chambered heart, with the upper left and right chambers serving as reservoirs for blood returning to the heart and the lower ones serving as pumping devices for blood leaving the heart. All mammals also have a similar arrangement of arteries and veins directing blood into and out of the heart. More quantitatively, as we shall see in the next section, the weight of the heart and the weight of the blood in the system are both directly proportional to the weight of the mammal. Similarity of operation is likewise revealed by consideration of resting heart rate, average blood flow out of the heart per minute, and other such variables. As will be seen, resting heart rate and blood flow are also dependent only on mammal size and can be predicted from simple algebraic equations involving only the mammal weight.

Such similarity indicates that the design of the cardiovascular system of all mammals is governed by the same basic design theory. The term design, as used here and throughout this work, should be considered to indicate simply the makeup of the system as we find it from experimental observations. Likewise, design theory should be taken to mean the fundamental principles governing the design and operation of the system. When viewed as an engineering system in the manner described earlier, there are plainly a number of significant design variables associated with it, among them being, for any given mammal, the pump size, the pump rate, the volume output, the piping dimensions, and the characteristics of the diffusion system. The design theory revealing the magnitude and interrelation of these and other variables for a particular animal size provides an intriguing study in both engineering and physiology and forms the subject of the present book.

1.1 SCALING LAWS

Central to our study will be the various scaling laws that have been found to apply to physiological variables of mammals ranging in size from mice to elephants. These laws relate parameters such as heart weight and heart rate to animal weight and give us direct information about certain design values for any given size mammal. As part of our general study of the design of the cardiovascular system, we will find it useful to examine here a few of the more common relations known to exist among mammals of widely differing size. In later chapters, we will consider the origin of these as well as other scaling relations in connection with a specific design theory.

Scaling laws for animals are generally of the power-law form such that a measured variable, say Y, is related to the animal weight, say W, by the relation

$$Y = K_c W^n \tag{1.1}$$

where K_c is a coefficient and n is the exponent of the power law. In applying this law to specific data, it is convenient to write it in its logarithmic form:

$$\log Y = n \log W + \log K_c \qquad (1.2)$$

so that a plot of $\log Y$ against $\log W$ on linear-scale axes, or a plot of Y against W on logarithmic-scale axes, will yield a straight line of slope n and intercept $\log K_c$ at $\log W = 0$, or K_c at $W = 1$. Appendix A illustrates these properties in greater detail. In biology and physiology, relations of the form of Eq. (1.1) are often referred to as allometric equations in keeping with terminology originally proposed by Huxley (1932).

For our first description of such scaling laws, we consider typical data on the empty heart weight w_H of various mammals, as listed in Table 1.1. If we assume a governing relation of the form of Eq. (1.2) and apply best-fit methods of analysis to the heart–weight data, we find the equation in the logarithmic, base 10, form expressible as

$$\log w_H = 1.02 \log W - 2.36$$

In the power law form of Eq. (1.1), the exponent n is equal to 1.02 and the coefficient K_c is equal to 0.0043, since log (0.0043) is equal to –2.36. The accuracy and range of the data in Table 1.1 are, however, not adequate to justify the value of 1.02 over the rounded value of 1.0 for the value of n. The equation for heart weight may thus be written in the form of Eq. (1.1) simply as the proportional relation

$$w_H = 0.0043\, W \qquad (1.3)$$

For a second illustration, we consider typical data on total blood weight w_B, also tabulated in Table 1.1. In the same manner as above, we may calculate the best-fit relation for total blood weight as

$$\log w_B = 0.98 \log W - 1.25$$

Again, on rounding the value of 0.98 to unity, we may express this result in the form of Eq. (1.1) as the simple proportional relation

$$w_B = 0.056\, W \qquad (1.4)$$

It is of interest to see how well these equations describe the actual measurements. The data in Table 1.1 have accordingly been plotted in Fig. 1.1 using graphs with logarithmic-scale axes, as discussed earlier in connection with the general form of Eq. (1.2). Also shown are predictions from Eqs. (1.3) and (1.4). We see from the comparison that the agreement is, indeed, remarkably good over the entire rat-to-steer range of measurements.

Equations (1.3) and (1.4) are simple proportional relations showing that heart weight and total blood weight scale directly with animal weight for any given mammal. Not all animal scaling laws are, however, simple

TABLE 1.1.   Typical Data on Heart Weight and Total Blood Weight
             for Various Mammals

Animal	Body weight (kg)	Heart weight (kg)	Blood weight (kg)
Rat	2.5×10^{-1}	9.4×10^{-4}	1.5×10^{-2}
Guinea pig	9.0×10^{-1}	2.3×10^{-3}	2.8×10^{-2}
Monkey	4.5	2.3×10^{-2}	3.0×10^{-1}
Dog	1.0×10^{1}	8.5×10^{-2}	7.0×10^{-1}
Human	6.0×10^{1}	3.2×10^{-1}	4.3
Steer	7.0×10^{2}	2.3	2.5×10^{1}

Data source: Brody (1945).

proportional relations, as our next important illustrations show. We consider, in particular, data on resting heart rate (or pulse rate) and oxygen-consumption rate of mammals, as listed in Table 1.2.

Using best-fit methods of analysis, as before, we find the heart rate ω expressible in the form of Eq. (1.2) as

$$\log \omega = -0.259 \log W + 2.36$$

Now, consistent with the accuracy of the data, we may replace the value of 0.259 by 0.25 and express the heart rate in alternate form as the reciprocal one-fourth relation:

$$\omega = 229 \, W^{-1/4} \tag{1.5}$$

TABLE 1.2.   Heart Rate and Oxygen-Consumption Rate of Various Mammals

Animal	Body weight (kg)	Heart rate (beats/min.)	Oxygen consumption (ml/min.)
Mouse	2.5×10^{-2}	600	1.45
Rat	2.0×10^{-1}	440	6.57
Guinea pig	4.0×10^{-1}	267	5.44
Rabbit	2.0	205	1.50×10^{1}
Dog	6.5	120	6.22×10^{1}
Sheep	5.0×10^{1}	75	1.98×10^{2}
Human	6.5×10^{1}	60	2.37×10^{2}
Steer	7.3×10^{2}	48	1.61×10^{3}
Elephant	3.7×10^{3}	31	9.38×10^{3}

Data source: Brody (1945).

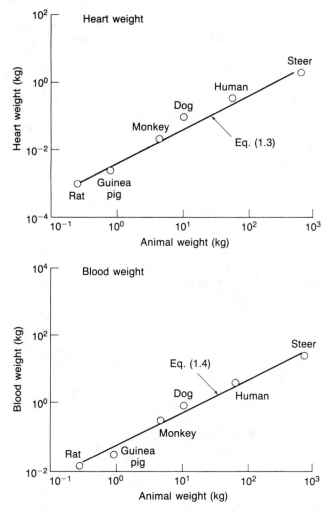

Fig. 1.1 Comparison of measurements of heart weight and blood weight given in Table 1.1 with predictions from best-fit Eqs. (1.3) and (1.4), respectively.

The data on oxygen-consumption rate Q_o likewise provide a best-fit relation of the form

$$\log Q_o = 0.758 \log W + 1.05$$

which may be rounded and expressed as the three-fourths relation:

$$Q_o = 11.2 \, W^{3/4} \tag{1.6}$$

To see how well these equations apply, we may plot the data given in

Table 1.2 and compare them with predictions from the derived equations. This has been done in Fig. 1.2 using graphs with logarithmic-scale axes, as before. We see again a close correlation between the individual data and the best-fit equations, in this case for the mouse-to-elephant range of measurements.

Having the above scaling law for oxygen-consumption rate, we may also establish the law for heat production of a resting mammal, since 4.8 kcal is normally considered the metabolic heat equivalent of one liter of oxygen. Hence, heat production Q_h in units of kcal/min. is expressible as

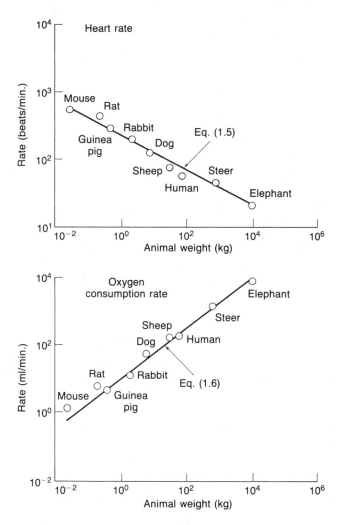

Fig. 1.2 Comparison of measurements of heart rate and oxygen-consumption rate given in Table 1.2 with predictions from best-fit Eqs. (1.5) and (1.6), respectively

$$Q_h = 0.0538 \ W^{3/4} \tag{1.7}$$

In a similar way, it is known that about 5 ml of oxygen are delivered by every 100 ml of blood during a cycle of circulation in a resting mammal, so that cardiac output Q_b, or average blood flow, in units of ml/min. can be expected to be about 20 times the average rate of oxygen consumption, that is

$$Q_b = 224 \ W^{3/4} \tag{1.8}$$

In connection with these last equations, we may note that for a human weighing 70 kg, the resting heat production is typically in the range of 1.2 kcal/min.to 1.3 kcal/min., and the cardiac output is typically in the range of 5000 ml/min. to 5500 ml /min. Equations (1.7) and (1.8) give corresponding values of 1.30 kcal/min. and 5420 ml/min., respectively, thus providing close agreement with typical measurements.

After collecting all of the above results, we see that heart weight and blood weight scale directly with animal weight, that resting heart rate scales with animal weight by a reciprocal one-fourth power law, and that resting oxygen-consumption rate, heat production, and cardiac output scale by a three-fourth power law.

As indicated earlier, these empirical scaling relations have been known for many years. The proportional scaling laws for heart and blood weights were stated more than 40 years ago by Brody (1945). Some 20 years earlier, the reciprocal one-fourth law for heart rate was essentially given by Clark (1927), who reported the exponent as –0.27 rather than the –0.25 used here. At about the same time, the three-fourths law for oxygen-consumption rate and its related metabolic heat-production measure was specifically stated by Kleiber (1932, 1961), with Brody and his co-workers (Brody 1945) independently proposing a nearly identical power law with exponent 0.73.

1.2 IMPLICATIONS TO DESIGN

The existence of the above scaling laws implies a fundamental similarity in the design (or makeup) of the cardiovascular system of mammals such that no one species is favored over another in providing for its basic physiological needs. The cardiac output of the human, for example, is the same, on the average, as that for sheep or pigs of the same weight. It is also between that of the dog and horse, consistent with the relative sizes of these animals. If we suppose that the cardiovascular system for the human is best designed for its size, we may equally well suppose, on the basis of the similarity of the system, that those for other mammals are also best designed for their respective sizes.

In engineering terms, we may, as noted earlier, interpret this similarity in the design of the cardiovascular system as meaning that all systems are governed by the same design theory and related design equations. When we study the variation of physiological parameters among mammals, we are therefore studying the influence of an independent variable in these equations, the animal size or weight, on a dependent variable such as cardiac output.

In this connection, we may make an interesting well-known observation about the oxygen-consumption rate of mammals when expressed in terms of unit mammal weight. Specifically, from the data in Table 1.2, we may observe that the oxygen-consumption rate for one gram of average mouse tissue is 1.45/25, or 0.058 ml/min., while that for, say, one gram of average rabbit tissue is 15/2000, or 0.0075 ml/min. Thus, we see that one gram of average mouse tissue consumes oxygen at a rate nearly eight times greater than that of one gram of average rabbit tissue. More generally, as can be seen by dividing both sides of the scaling relation of Eq. (1.6) by the animal weight, the smaller the mammal, the higher its rate of oxygen consumption per unit of its weight. From the scaling relations of Eqs. (1.7) and (1.8), the same is also seen to be true for the heat produced by a gram of average tissue and for the fraction of cardiac output delivered to it. At first sight, these results are perhaps not those expected, since one gram of average tissue might be thought to operate in the same manner for all mammals. For this to be the case, however, the scaling Eqs. (1.6) through (1.8) would have to be simple proportional relations rather than the three-fourths power relations that the measurements indicate.

While the rate of oxygen consumption of one gram of average tissue varies with mammal size, there does exist invariant behavior with respect to the amount of oxygen consumed over a time interval proportional to the time between heartbeats. This is readily seen by dividing the oxygen-consumption rate given by Eq. (1.6) by the animal weight and then multiplying by the time between heartbeats. Since this time is given by the reciprocal of the heart rate, the latter operation is equivalent to dividing by the heart rate, as expressed by Eq. (1.5). The result is

$$\frac{Q_o}{\omega W} = 0.049 \frac{\text{ml}}{\text{kg-cycle}}$$

which is, indeed, independent of animal weight. The implication is that the utilization of oxygen by, say, one gram of average tissue during the period of a heart cycle (or some fraction or multiple of it) is the same for all mammals, and that it is only the rate at which the tissue operates that varies with animal size. A more subtle implication is that a gram of average tissue utilizes oxygen in a periodic, rather than uniform, manner with periodic rate proportional to heart rate.

As a further illustration of the scaling relations of the previous section, we may consider the design of a large horse weighing ten times that of the human. We then see from these relations that its heart and total amount of blood should each weigh ten times as much as those of the human, that the rate of heartbeat of the horse should only be about one-half that of the human, and that the cardiac output should be about five times that of the human. If we notice that the ratio of cardiac output (average blood flow per minute) to heart rate (beats per minute) equals the volume of blood pumped by the heart per beat, we also see for the horse that this is ten times that of the human, in conformity with the relative weights of the two.

But why the reduced heart rate and cardiac output? This last result could also have been achieved by leaving the heart rate unchanged and having the cardiac output vary directly with animal weight. As noted above, this would then make the fraction of blood flow, oxygen consumption rate, and heat production for a unit weight of average tissue the same for both horse and man. The answer to the question is that factors other than bulk size must enter into the overall design considerations. Past attempts to identify these constraints have involved mainly considerations based on heat loss and heat production of mammals, or considerations based on dimensional analysis of some of the main physiological variables involved. The first mentioned studies are associated with the work of Sarrus and Rameaux in 1837, and the second with that of Lambert and Teisser in 1927. Both ultimately fail to identify proper constraints on the cardiovascular system, but are of interest in providing perspective for the present work.

1.3 HEAT PRODUCTION THEORY

That constraints other than size enter into animal design was recognized as long ago as 1837 by French scientists Sarrus and Rameaux (Brody 1945), who argued that heat loss to the environment by an animal must be proportional to body surface area and, hence, that heat production by the animal must also be proportional to body surface area if body temperature is to be maintained. They also argued that heat production must be proportional to the rate of oxygen consumption and cardiac output so that these last variables must likewise vary with surface area of the body.

Now, as described in detail in Appendix B, the surface area of each of a family of similarly shaped bodies of different size will be proportional to the square of any linear dimension of the body, while the volumes will be proportional to the cube of any linear dimension. Moreover, for bodies made of the same material, their weights will be proportional to their volumes so that we have the converse result that the linear dimensions of the bodies must be proportional to the cube root of their weights. Their surface areas will therefore vary with body weight raised to the two-thirds power. In this way, we see that, according to the reasoning of Sarrus and Rameaux, the resting heat production of mammals should all vary with animal weight by a two-thirds power law. Since the oxygen-consumption rate and cardiac output are proportional to heat production, these should likewise follow the two-thirds law of this theory. If, as noted earlier, blood volume pumped per beat by the heart is proportional to animal weight, it also follows that heart rate should vary with animal weight by a reciprocal one-third power-law relation.

These relations are not those now known to apply to mammals, although they do not differ greatly from the empirical relations described earlier for heat production, oxygen consumption, and cardiac output (three-fourths relation) and for heart rate (reciprocal one-fourth relation). The difference is, however, real and significant, as emphasized by Kleiber (1932) and Brody and co-workers (Brody 1945). The distinction is illustrated in Fig. 1.3, where Kleiber's original data on resting heat production are plotted against animal weight for the rat-to-steer range of mammals considered by him. Also plotted are measurements of Benedict (1938) and

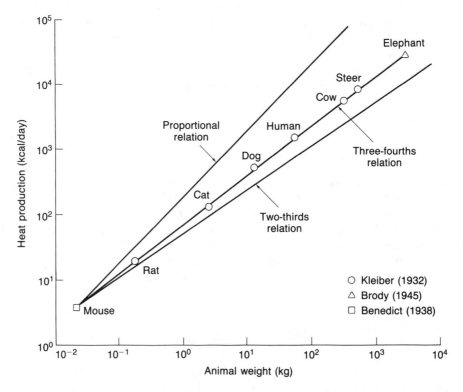

Fig. 1.3 Comparison of proportional, three-fourths, and two-thirds power-law relations with measurements of heat production of various mamals

Brody (1945) for the mouse and elephant, respectively. These data are further compared in Fig. 1.3 with predictions from a proportional relation, a two-thirds relation, and the three-fourths relation of Eq. (1.7), derived earlier from the completely independent measurements in Table 1.2. It can be seen that the three-fourths relation provides remarkably close predictions of the data, in contrast with both the proportional and the two-thirds relations.

Of course, without further study, the possibility must be admitted that the above difference between measurement and the two-thirds relation is due not to a fundamental problem with the reasoning of Sarrus and Rameaux but rather simply to a departure from the rule relating surface area to body weight raised to the two-thirds power. It will be recalled that this rule is exact for geometrically similar bodies made of the same material, but not necessarily so for differently shaped ones. Most mammals do, of course, have approximately similar shapes, and this provides some justification for the surface rule. Fortunately, though, extensive measurements of the surface areas of various mammals have been made which allow a direct examination. Table 1.3 lists some typical measurements.

Assuming a relation of the form of Eq. (1.2) between surface area S_a and animal weight W, we find from best-fit methods of analysis that the data in Table 1.3 give the relation

$$\log S_a = 0.66 \log W - 0.986$$

which does, indeed, correspond to a two-thirds scaling relation in the form

$$S_a = 0.103 \, W^{2/3} \tag{1.9}$$

The data in Table 1.3 are plotted in Fig. 1.4 and compared with the predictions from Eq. (1.9). We see that the agreement is consistently good throughout the entire rat-to-cow range of data. Thus, even though the shape of mammals such as rats, dogs, and cows are only approximately

---

TABLE 1.3   Typical Measurements of Body Surface Area of Mammals

Animal	Body weight (kg)	Surface area (m^2)
Rat	2.00×10^{-1}	3.01×10^{-2}
Dog	3.10	2.42×10^{-1}
Dog	1.82×10^{1}	7.66×10^{-1}
Dog	3.12×10^{1}	1.08
Human	7.00×10^{1}	1.96
Cow	5.00×10^{2}	4.87

Data source: Brody (1945) for rat, human, cow; Schmidt-Nielsen (1972) for dogs.

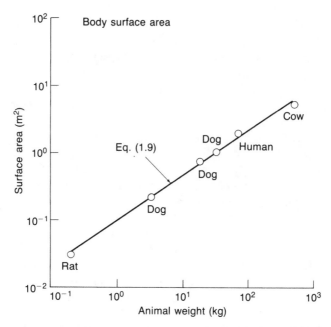

Fig. 1.4 Comparison of measurements of body-surface area in Table 1.3 with predictions from the best-fit Eq. (1.9).

similar to one another, their surface areas vary with body weight in the same way as if this were precisely the case.

With this last result, we now see that the theory of Sarrus and Rameaux relating heat production directly to the surface area of the body cannot be viewed as fundamentally correct, with variations from measurement attributed simply to the area–weight relation of mammals. The latter does not obey a three-fourths power law, as would be required for agreement of this theory with heat-production measurements, but instead obeys a two-thirds law like that holding for similarly shaped bodies made of identical materials.

This observation does not, of course, contradict the basic heat-balance argument of Sarrus and Rameaux as related to body temperature. In the case of resting mammals, the ratio of heat produced to heat lost must, indeed, be independent of animal size so that, consistent with observation, an essentially identical body temperature is achieved for all mammals. The conclusion from the above is therefore that heat loss of mammals is proportional, not to surface area, but instead to animal weight raised to the three-fourths power. This suggests, in turn, that the heat loss is an adaptive consequence of the heat production rather than a fundamental reason for it.

One final comment that bears on the heat-production theory should also be made. This concerns the oxygen-consumption rates of cold-blooded animals which, although considerably lower than mammals, have also been found to obey non-proportional relations with animal weight (Schmidt-Nielsen 1972, 1984). Such animals have, of course, no reason to consume oxygen and produce heat in proportion to their surface area, or any approximation to it, their body temperature being that of the surroundings. Thus, apart from differences between predicted and measured power-law relations, the existence of non-proportional relations for oxygen-consumption rate for both warm-blooded and cold-blooded animals further suggests that their origin lies in fundamental physiological processes rather than in heat-balance requirements.

1.4 DIMENSIONAL ANALYSIS

Attempts to use dimensional analysis and simple modeling theory to establish scaling laws and design restraints for the cardiovascular system date from the work of Lambert and Teisser in 1927 (Brody 1945; Kenner 1972). In modern engineering practice, this procedure involves identifying a quantity, or variable, of interest in a system and subsequently listing all the various quantities and parameters of the system on which it depends. The physical dimensions, or units, of each are next established and independent dimensionless ratios of the quantities and parameters formed. Because relationships between physical variables must be expressible in a manner independent of the units used to measure them, the ratio involving the quantity of interest may then be regarded as varying with the remaining ratios. Thus, when the latter are all held fixed under change of scale, the former will remain fixed also. Scaling requirements can therefore be investigated by considering changes in the geometric dimensions of the system while maintaining all the dimensionless ratios constant. Limitations on the method arise when conflicting conditions are required in order to maintain the ratios constant. In this case, no general scaling laws exist. Without additional knowledge of any groupings of the quantities and parameters of the system, it is also necessary that all length dimensions involved in describing the system geometry vary in the same way under change of scale. Further details of the method are described in Appendix C.

Let us now apply this procedure to the cardiovascular system. We may identify the cardiac output as the main quantity of interest. The associated governing quantities and parameters can be established by considering an idealized version of a pumping cycle of the heart. At the onset of pumping, the cardiac muscle can be envisioned, as a result of stimulation, to undergo a physiological change that causes the natural free length of the muscle to become less than its resting length. At this instant, the muscle thus appears in a stretched state and begins to contract under the action of elastic restoring forces. As it does, it must, of course, force the heart itself to contract and expel blood into the vascular system, thereby giving rise to the cardiac output.

With this picture in mind, suppose that we now let ℓ_1, ℓ_2, etc., represent the various length dimensions of the system such as mean heart diameter, lengths of arteries, diameters of arteries, and so on. The cardiac output can then be expected to depend on the geometry of the system as represented by ℓ_1, ℓ_2, etc., as well as on quantities representing the important physical aspects of the system. These include the inertial (acceleration) and frictional (velocity) resistances to blood flow in the vascular system, the contractile force and elasticity of the heart muscle, and the frequency of heartbeat. The appropriate characterization of these quantities in specific terms can be made by imagining a full mathematical expression relating the cardiac output Q_b (in units of volume per unit time) to the various independent parameters governing the description. These parameters will be discussed in detail in Chapter 2 in order to establish a basis for a suitable design theory. For our present simpler needs, we may, however, identify them using basic physical arguments. They are also discussed in general terms in Appendix D.

In this way, we observe that the mass density ϱ of the blood (in units of mass per unit volume) must appear in the mathematical expression for cardiac output as the significant parameter associated with the inertial resistance of the blood flow. In the same way, we also see that the viscosity coefficient μ of the blood (in units of pressure times time) must appear as the significant parameter insofar as frictional resistance of blood flow is concerned, this coefficient being the material constant relating resistive force to flow for a given fluid. The contractile periodic force of the heart may next be represented in our mathematical expression by its amplitude F_o (in units of force) and its rate ω (in units of reciprocal time). Finally, the elasticity of the heart may be represented by the elastic modulus E of the cardiac muscle (in units of force per unit area), this being the material constant relating force and extension in a given solid.

Upon assembling the above results, we accordingly have the expression for the cardiac output defined in general function form as

$$Q_b = Q_b\,(\varrho\,,\mu\,,F_o\,,E,\omega\,,\ell_1,\ell_2) \tag{1.10}$$

where $Q_b\,(-)$ denotes a function of the indicated variables and where, for sake of brevity, we have listed only two of the many length dimensions involved. Now, if F, M, L, and T denote the physical units of force, mass, length, and time, respectively, that are used to measure the variables in Eq. (1.10), we may express the individual units of the variables in the manner indicated in Table 1.4.

TABLE 1.4.   Physical Dimensions, or Units of Variables in Eq. (1.10)

Variable	Symbol	Units
Cardiac output	Q_b	L^3/T
Density	ϱ	M/L^3
Viscosity	μ	FT/L^2
Cardiac force	F_o	F
Elastic modulus	E	F/L^2
Heart rate	ω	$1/T$
System lengths	ℓ_1, ℓ_2	L

Observing from Newton's Second Law of Motion that the unit of mass is connected to that of force through the relation $M = FT^2/L$, we may easily form the following independent dimensionless (or non-unit) groups from the variables in Eq. (1.10):

$$\frac{Q_b}{\omega \ell_1^3}, \frac{\varrho \omega^2 \ell_1^2}{E}, \frac{\mu \omega}{E}, \frac{F_o}{E \ell_1^2}, \frac{\ell_2}{\ell_1}$$

Since Eq. (1.10) must be expressible in dimensionless form in order to make it independent of the choice of the units used to measure the variables, we may therefore rewrite it in terms of the above dimensionless variables as

$$\frac{Q_b}{\omega \ell_1^3} = f\left(\frac{\varrho \omega^2 \ell_1^2}{E}, \frac{\mu \omega}{E}, \frac{F_o}{E \ell_1^2}, \frac{\ell_2}{\ell_1} \right) \tag{1.11}$$

where $f(-)$ denotes a function of the indicated variables.

From this last equation, we see that if all the independent variables on the right-hand side of the relation are constant for different scale (or size) systems, the dependent variable on the left-hand side must also be constant for the different scale systems. Let us now examine this possibility. We assume at the outset that the density ϱ and coefficient of viscosity μ of the blood and the elastic modulus E of the heart muscle are all constants for different size mammals, consistent with observation. The last ratio ℓ_2/ℓ_1 on the right-hand side of Eq. (1.11) is, of course, constant for uniform change of scale where all linear dimensions are changed by the same factor. Similarly, the ratio $F_o/E \ell_1^2$ will be constant for different scale systems provided the amplitude of the contractile force of the heart varies as the square of any linear dimension of the system. However, on considering the first two ratios $\varrho \omega^2 \ell_1^2/E$ and $\mu \omega/E$, we see that, for constant values, the first requires the heart rate to vary inversely with any linear dimension of the system, while the second requires it to be constant for different scale systems. Thus, it is impossible to have all the independent variables of Eq. (1.11) constant for different size systems and hence to infer any scaling relation for cardiac output from this general theory and the assumption of uniform scaling.

If the viscous effects are completely ignored, contrary to their importance in the cardiovascular system, scaling relations can then be developed from Eq. (1.11) by neglecting the dimensionless variable containing the viscosity coefficient. This approach leads to relations identical to those in the Lambert-Teisser scaling scheme for biological systems as described by Kenner (1972). Apart from being founded on an incorrect simplification, this routine gives scaling results that do not generally fit the facts except in an approximate way. Thus, it requires that heart rate scale inversely with any length dimension of the system and that cardiac output vary with the square of any length dimension. As noted earlier, for uniform change of scale of systems made out of the same materials, the linear dimensions must be proportional to the cube root of the system weight. Thus, we see that the neglect of blood viscosity in Eq. (1.11) leads to the incorrect scaling results that heart rate must vary with animal weight by a reciprocal one-third power law and that cardiac output must vary with animal weight by a two-thirds power law.

It may be recognized that these results are the same as those arrived at earlier using heat-production and surface-area arguments of Sarrus and Rameaux. In those arguments, no consideration was given to whether such scalings were mechanically possible. We now see that, with viscous resistance to blood flow included and the assumption of uniform scaling, they are not.

An alternate, perhaps more reasonable, simplification of Eq. (1.11) that allows extraction of scaling relations might be to assume that inertial resistance is negligible in comparison with viscous resistance. In this case, Eq. (1.11) then yields scaling relations requiring the heart rate of mammals to remain constant under change of scale and requiring cardiac output to vary directly with animal weight. These results are, of course, also inconsistent with observations.

Our conclusion from the above discussion is therefore that dimensional reasoning alone cannot yield accurate scaling laws and design constraints for the cardiovascular system of animals. Such does not, of course, imply that the experimental facts are beyond the reach of engineering theory and the general laws of mechanics. Rather, it means simply that, for a full understanding of the design constraints and scaling relations, a more specific design theory must be used than that based solely on dimensions and the attendant requirement of uniform scaling.

In connection with this last remark, we may note the interesting paper by McMahon (1973) in which non-uniform scaling is considered in biological problems. Using arguments based on elastic buckling and bending of beams, he first showed for similar response characteristics of cylindrical bodies that lengths must vary with diameters raised to the two-thirds power. It was next observed that this same elastic scaling requires the cross-sectional area of cylindrical bodies to vary with their weight raised to the three-fourths power. Taking the argument over to animals and assuming (a) that this geometry applies to muscles, (b) that their cross-sectional areas are proportional to maximal rate of oxygen consumption, or heat generation, and (c) that maximal and resting values are proportional, the three-fourths scaling law for oxygen-consumption rate was then established as a consequence of the elastic scaling law (see also McMahon and Bonner 1983).

This idea is interesting because it leads to the right answer. However, it is based on the above three broad assumptions, as well as the fundamental assumption that elastic scaling of beams plays a central role in the process. Thus, while worthy of consideration, the work does not reveal how the three-fourths scaling law for oxygen-consumption rate, and associated cardiac output, is possible in terms of the basic operations of the caviovascular system.

1.5 SUMMARY

We have seen that certain well-known scaling results for the cardiovascular system of mammals cannot be explained by classical arguments involving heat production and body surface area, nor can they be explained using conventional modeling theory and the attendant assumption of uniform scaling of all linear dimensions. With these known scaling laws, as well as others not yet discussed, we thus find ourselves in the position of having before us a well-organized body of empirical information for which no consolidating theory is currently available.

In the present book, we accordingly study the design of the cardiovascular system in some detail and develop an engineering theory that is indeed capable of predicting the various scaling laws known to apply. In this way, we shall then establish the origin of these laws and, more importantly, establish some new concepts about the design and workings of the system. Our attention will be directed mainly toward mammals. However, many of the results established will be general enough to apply also to other animals having similar, though perhaps simpler, cardiovascular systems.

In Chapter 2, we discuss certain features of the cardiovascular system, including its organization and its engineering-like components. These are subsequently used in Chapter 3 to construct a detailed theory of the system which allows interrelation of significant parameters and the derivation of basic scaling laws. In Chapter 4, we demonstrate the adequacy of these and additional derived scaling laws using measured physiological parameters from mammals. This study is continued in Chapter 5 when we consider aspects of the individual organs of the body. In Chapter 6, we examine specific response characteristics of the cardiovascular system using a simplified version of the theory established in Chapter 3. Finally, in Chapter 7, we summarize the results of our study and emphasize important findings relative to the engineering design of the system.

REFERENCES

BENEDICT, F. G. 1938. *Vital Energetics: A Study in Comparative Basal Metabolism*. Washington, DC: Carnegie Institute of Washington.

BRODY, S. 1945. *Bioenergetics and Growth, With Special Reference to the Efficiency Complex in Domestic Animals*. New York: Reinhold Publishing.

CLARK, A. J. 1927. *Comparative Physiology of the Heart*. Cambridge: Cambridge University Press.

HUXLEY, J. S. 1932. *Problems of Relative Growth*. London: Methuen.

KENNER, T. 1972. Flow and Pressure in the Arteries. In *Biomechanics—Its Foundations and Objectives,* Y. C. Fung, N. Perrone, M. Anliker, eds., pp. 381–434. Englewood Cliffs: Prentice-Hall.

KLEIBER, M. 1932. Body Size and Metabolism. *Hilgardia* 6: 315–53.

———. 1961. *The Fire of Life. An Introduction to Animal Energetics*. New York: John Wiley and Sons.

MCMAHON, T. 1973. Size and shape in biology. *Science* 179: 1201–4.

MCMAHON, T., and J. T. BONNER, 1983. *On Size and Life*. New York: Scientific American Books.

SCHMIDT-NIELSEN, K. 1972. *How Animals Work*. Cambridge: Cambridge University Press.

———. 1984. *Scaling: Why is Animal Size So Important?* Cambridge: Cambridge University Press.

2

General Features of the Cardiovascular System

In order to construct a useful engineering design theory for the cardiovascular system of mammals, we need to consider both the overall organization of the system as well as its main components: the blood, the heart, and the vascular system. This we do in the present chapter. Our interest will be in reviewing the basic operation of the system and in developing appropriate engineering descriptions of its essential parts.

2.1 ORGANIZATION

The general organization of the cardiovascular system of mammals is well known. Our basic understanding dates from the time of William Harvey (1578–1657) and his discovery of the circulation of the blood. In modern engineering terms, the essential features of the cardiovascular system may be illustrated as in Fig. 2.1 (see, for example, Noordergraaf 1978). The system is seen to consist of two pumps in series, the left and right sides of the heart, together with connecting pipes, the arteries and veins, and various diffusion devices, the capillaries. Blood flow from the left side is to the systemic vascular bed. It is directed first through the aorta and then through a series of branching arteries to the capillaries of the various parts of the body. After exchange of products in these parts by diffusion, filtration, and absorption processes, the blood is collected by a series of coalescing veins and returned to the right side of the heart. From here, the blood is pumped to the pulmonary vascular bed. The blood flow is thus

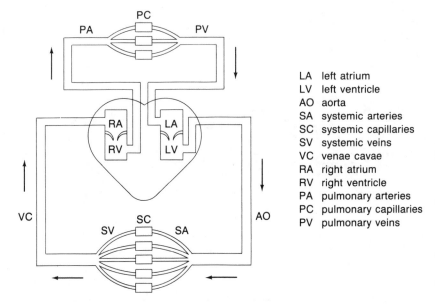

LA left atrium
LV left ventricle
AO aorta
SA systemic arteries
SC systemic capillaries
SV systemic veins
VC venae cavae
RA right atrium
RV right ventricle
PA pulmonary arteries
PC pulmonary capillaries
PV pulmonary veins

Fig. 2.1 Illustration of overall organization of the cardiovascular system of mammals.

directed through the pulmonary artery and through a series of branching arteries to the capillaries of the lungs where gas exchange takes place. The enriched blood is finally returned through a coalescing venous system to the left side of the heart for recirculation.

The working fluid of the system, the blood, consists of a suspension of components in an aqueous solution, so designed to carry out the transport functions required of it. The flow characteristics of the blood are also consistent with the available power of the heart for circulating the blood throughout the body.

The system pumps, the left and right sides of the heart, perform their difficult tasks by pulsatile action. In the human, for example, the heart squeezes blood into the system at a rate of about 70 beats per minute. For smaller mammals, the rate is greater, and for larger mammals, it is less. Thus, as we saw in Chapter 1, the heart rate of the mouse is about 600 beats per minute, while that of the elephant is only about 30 beats per minute. Each side of the heart, that is, each pump, consists of two chambers, an atrium and a ventricle. The left atrium collects blood returning in the veins from the lungs, and the left ventricle pumps this blood to the body. The right atrium likewise collects the blood returning from the body, and the right ventricle pumps this blood to the lungs.

The resistance to blood flow is much greater in the systemic circulation than in that of the lungs, so that the left side of the heart operates at a considerably greater pressure than the right side. For the left side, the maximum and minimum pressures in the aorta of the human (and other mammals) are normally about 120 mm Hg and 80 mm Hg, respectively. For the right side, the corresponding pressures in the pulmonary artery are about 25 mm Hg and 10 mm Hg, respectively.

The average blood flow to the body from the left side of the heart is approximately equal, within one or two percent, to that to the lungs from the right side. The flows would be identical except that a small amount of blood from the left side is directed to the bronchial vascular system of the lungs for tissue nourishment and then returned directly to the left side for further circulation through the systemic system. As noted in Chapter 1, this average flow, strictly from the left side, is referred to as the cardiac output. In the human, it is about 5 liters per minute, that is, about 80 gallons per hour. In contrast, in the elephant it is about 90 liters per minute, which is about 1400 gallons per hour.

The piping network of the system, the systemic and pulmonary vascular beds, consists of a complex assemblage of vessels designed to direct blood flow to the tissues for transfer of various substances and to the lungs for gas exchange. In each bed, the arteries, carrying blood from the heart, branch into larger and larger numbers of smaller and smaller vessels until at their finest subdivision there are, in the human, many hundreds of millions of these with diameters of the order of one one-hundreths of a millimeter and lengths of the order of one millimeter. These small vessels feed the capillaries where exchange between blood and the surroundings take place by diffusion, filtration, and absorption processes. The dimensions of the capillaries are of the same order as the very small arteries supplying them, but their number is greater, being estimated to be of the order of billions. On the venous side, the very small veins draining the capillaries are of the same general number and dimensions as the very small arteries feeding them. Likewise, the coalescing sets of veins returning blood to the heart roughly parallel in number and dimensions the corresponding sets of arteries at each level of division.

Interestingly, not all capillaries are open and functional in the resting state because the arteries connecting these inactive ones are constricted to prevent flow. This provides a reserve for the active state when increased needs of the cells arise and the constricted arteries respond by opening and allowing flow through the additional capillaries.

2.2 THE WORKING FLUID—THE BLOOD

We have already noted that the blood can be regarded in engineering terms as the working fluid of the cardiovascular system. As such, it is required to perform a number of important functions. Among them, for ex-

TABLE 2.1.   Main Components of Blood

Formed Elements

Component	Amount (human)
red cell	$4.5\text{–}6 \times 10^6/\text{mm}^3$
white cell	$5\text{–}8 \times 10^3/\text{mm}^3$
platelets	$2.5\text{–}5 \times 10^5/\text{mm}^3$

Plasma

Component		Amount (human)
water		91 ml/100 ml
fibrinogen	⎫	0.2–0.4 g/100 ml
globulin	⎬ Proteins	2.1–3.3 g/100 ml
albumin	⎭	3.5–5.3 g/100 ml

Additional Constituents

Inorganic	Organic	
sodium	urea	⎫ waste products
chlorine	uric acid	⎬ of metabolism
potassium	creatinine	⎭
calcium	amino acids	⎫
magnesium	fats	⎬ nutrients
phosphorus	glucose	⎭
carbon dioxide	others	
bicarbonate		
others		

Data source: Attinger (1973).

ample, are the transport of oxygen from the lungs to various parts of the body, the reverse transport of carbon dioxide from the cells to the lungs for discharge, the transport of nutrients from the digestive tract to the cells, the transport of waste products from the cells to the kidneys for discharge and the transport of hormones and other agents from manufacturing glands to the body cells for regulation. The blood also plays a major role in transferring heat from the warmer locations of the body where high metabolic activity exists to the cooler, less active regions, thereby assisting in the regulation of body temperature.

An analysis of the composition of the blood reveals that it consists of a suspension of formed elements (red cells, white cells, platelets) in a saline solution (plasma) containing proteins and various inorganic and organic constituents. Table 2.1 summarizes the major components in human blood.

Oxygen Transport and Exchange

Of the many requirements on the blood as a working substance, the most demanding is probably the transport of oxygen. In order to sustain mammalian life as we know it, the blood must leave the lungs carrying about 20 ml of oxygen in every 100 ml of blood. However, ordinary aqueous solutions will allow only a small fraction of this amount to be dissolved in them, so special arrangements must exist for this purpose. Nature has done this by including complex molecules of hemoglobin in the blood that provide the necessary transport function by chemical combination with the oxygen. These molecules are contained within the red blood cells (erythrocytes) which, together with the relatively fewer white blood cells (leukocytes) that are active in resisting and fighting disease, and the platelets (thrombocytes) that are active in blood clotting at damaged blood vessels, are suspended in the blood fluid, or plasma, and move with it as it circulates throughout the body.

In the normal human, there are approximately 5 million of the hemoglobin-rich red blood cells in each cubic millimeter of blood. These cells consist of a very thin flexible membrane enclosing a concentrated solution of the hemoglobin and are disc-like in shape with diameters of the order of the small capillaries through which they must pass. They account for about 40%–45% of the total volume of the blood (the hematocrit). The remainder is mainly the plasma, with only a small percentage occupied by the white blood cells and platelets.

Under the relatively high oxygen concentration found in the lungs, almost all of the hemoglobin circulating through these organs combines with oxygen to form oxyhemoglobin. Blood leaving the lungs is therefore almost fully saturated with oxygen, having the required level of about 20 ml of oxygen per 100 ml of blood that was noted earlier. As the blood circulates through the systemic tissues, the lower oxygen concentrations encountered cause the dissociation of oxygen from the oxyhemoglobin molecules and the discharge of oxygen to the cells by diffusion.

For a resting mammal, about 5 ml of oxygen are given up, on the average, to the systemic tissues for every 100 ml of blood flowing through them. Thus, on returning to the lungs, the mixed venous blood still contains about 15 ml of oxygen per 100 ml of blood, corresponding to about a 75% saturation level. Of course, this high level of reserve oxygen in the blood is available for use when increased activity requires it. The working fluid of the system is therefore designed to accommodate not only resting, or idling, needs of the body, but those arising from markedly increased output.

The amount of oxygen taken in or released from an elemental (or small representative) volume of blood during its passage through a capillary depends on the difference between the partial pressure of the oxygen in the blood and in the surrounding tissue. It will be recalled from chemistry that the partial pressure of a gas in a liquid in equilibrium can be thought of as the pressure exerted by the gas as it attempts to escape; or, equivalently, as the partial pressure of the gas surrounding the liquid and holding the dissolved gas in equilibrium. Thus, when the partial pressure of oxygen external to the elemental volume of blood is less than that of the blood itself, oxygen will be unloaded and the partial pressure of the oxygen in the blood will fall toward that of its surroundings. Similarly, when the external partial pressure of the oxygen is greater than that of the blood, oxygen will be absorbed and the oxygen partial pressure of the blood will rise toward the external value.

The process of oxygen transfer to the blood in a pulmonary capillary is illustrated schematically in Fig. 2.2 for the case of the human in the resting state. As indicated, an elemental volume of blood arrives at the capillary with an oxygen partial pressure Po_2 of 40 mm Hg. As it passes through the capillary, it picks up oxygen from the adjacent air in the lungs and its oxygen partial pressure comes to equilibrium with that of the inspired air, about 104 mm Hg. Because of the rapid diffusion of oxygen into the capillary, the elemental volume is seen to have reached oxygen equilibrium before it has traveled very far along the capillary. This provides a reserve capacity in the case of exercise where increased blood flow reduces the time that an elemental volume of blood is in the capillary.

The transfer of oxygen from an average systemic capillary of the human in the resting state is also illustrated schematically in Fig. 2.2. As in the case of the pulmonary capillaries, the rapidity of oxygen diffusion across the capillary wall allows an elemental volume of blood to come to equilibrium with the surroundings before it exits the capillary. For equilibrium in this case, oxygen must be unloaded from the blood to the surrounding interstitial fluid. Notice that the blood entering the systemic capillary is shown with an oxygen partial pressure of 95 mm Hg, rather than the 104 mm Hg that it had upon leaving the lungs. The difference is due mainly to the bronchial circulation, referred to earlier, which returns a small amount of venous blood directly to the left side of the heart.

For an engineering description of the above process, we may consider the net rate of transfer of oxygen from or to all the elemental volumes of blood instantaneously filling a capillary. The quantity to be considered is then the average rate of diffusion of oxygen across the capillary walls. This is known to be directly proportional to the difference ΔP_o between the average partial pressure of the oxygen in the capillary (average over the capillary length) and that in the surrounding interstitial fluid. The average diffusion rate is also known to be directly proportional to the lateral surface area A_c of the capillary and inversely proportional to the wall thickness h_c. Thus, the average volume of oxygen that moves

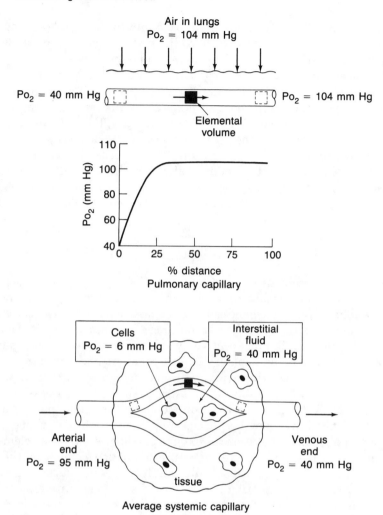

Fig. 2.2 Illustration of oxygen-exchange process in pulmonary and average systemic capillaries, with representative values of oxygen partial pressures for the human. Data source: Guyton (1971).

across the capillary wall per unit time, say J_G, can be expressed mathematically (see Appendix D) as

$$J_G = D_G \frac{\Delta P_o \, A_c}{h_c} \tag{2.1}$$

where D_G denotes a constant. When the pressure difference is positive, the rate of movement given by this equation is out of the capillary, and when the pressure difference is negative, the rate of movement is inward.

We may next use this equation to examine some simple scaling relations applicable to the process. Since the surface area of the capillary is proportional to the product of the radius and length, we see, for example, that, if all the dimensions of the capillary are doubled and the difference in partial pressure is held fixed, the diffusion rate will itself double. More generally, if the radial dimensions (radius and wall thickness) are varied in one manner and the length dimension in another, the diffusion rate will change as the product of the pressure difference and the capillary length changes. In addition, if there are a total of n_c such capillaries in the entire systemic vascular bed, the combined diffusion rate from all of these must equal the oxygen consumption rate Q_o discussed earlier in Chapter 1. In this case, the resulting scaling relation is then expressible as

$$Q_o \; \alpha \; \Delta P_o n_c L_c \tag{2.2}$$

where L_c denotes capillary length and α denotes proportionality under change of scale.

Since, as demonstrated in Chapter 1, the oxygen-consumption rate of mammals scales with body weight raised to the three-fourths power, the product on the right-hand side of this last relation must also scale in this manner. Insofar as this relation alone is concerned, this could be achieved from a design perspective by holding, say, the average partial-pressure difference and capillary length constant and varying only the number of capillaries. However, additional design requirements will be found in Chapters 3 and 4 that will require changes in all three of the variables of this product in order to accommodate different animal sizes.

As implied in our earlier discussion, there is a direct, though non-proportional, relation between the oxygen partial pressure in an elemental volume of blood and the degree of saturation of the element by the oxygen. Such a relation is established by simultaneous measurements of the partial pressure and relative saturation of a representative sample of blood. Figure 2.3 illustrates a typical so-called oxygen-dissociation curve derived from measurements of this kind. The main curve relating percent saturation to oxygen partial pressure is that for the case of normal arterial blood of the human, with pH of 7.4 and carbon dioxide partial pressure (Pco_2) of 40 mm Hg. Also indicated in general terms are the changes in the curve that result from decreasing and increasing carbon dioxide content, as measured by its partial pressures, and the changes that result when measurements are made on blood from larger and smaller mammals.

It can be seen in Fig. 2.3 that, for oxygen partial pressures in the range of 100 mm Hg to 110 mm Hg, the blood, or more precisely the hemoglobin, is almost completely saturated, irrespective of the particular carbon dioxide content or mammal size. Interestingly, the oxygen partial pressure for mixed air in the lungs of mammals is in this range, having been reduced from its atmospheric value of about 160 mm Hg because of

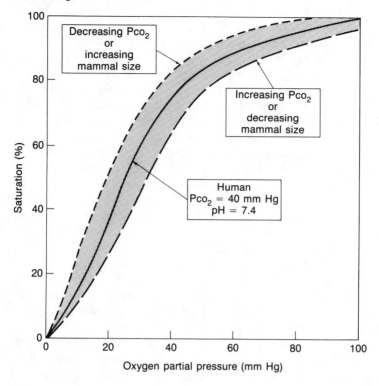

Fig. 2.3 Oxygen-dissociation curve for normal arterial blood of the human, with variable range, as indicated, for changing conditions. Source: Schmidt-Nielsen (1964) and Lloyd (1974).

mixing of newly inspired air with residual air, among other reasons. Thus, blood leaves the lungs with essentially a full load of oxygen, at an oxygen partial pressure of about 100 mm Hg, for transport to the systemic tissues. Some additional aspects of the characteristics indicated in Fig. 2.3 are discussed below.

From an engineering perspective, the oxygen-dissociation curve can be viewed as describing a significant property of the working fluid of the system. The detailed variation of percentage saturation with oxygen partial pressure, as indicated by the shape of the curve, reveals that the blood is indeed well suited to the purpose of oxygen uptake and unloading in the capillaries. The flatness of the curve in the region of 100% saturation means, for example, that the arterial oxygen content is relatively insensitive to the actual oxygen partial pressure of the air in the lungs. The blood can therefore arrive at the systemic tissues with a nearly constant amount of oxygen, even though the oxygen partial pressure in the lungs may vary somewhat. The relatively steep part of the curve in the venous range of

70%–75% saturation also has engineering significance. It means that, for oxygen transfer in the systemic capillaries, small variations in the oxygen partial pressure adjacent to the capillaries can regulate large differences in the amount of oxygen unloaded from the blood in order to meet individual tissue requirements. Thus, the oxygen partial pressure in the interstitial fluid surrounding the capillaries can be maintained at a sufficiently high level to ensure diffusion of oxygen onward to the cells even when oxygen demands of a particular tissue are high.

In addition to the importance attached to the shape of the standard oxygen-dissociation curve, the shift to the right, indicated in Fig. 2.3, for increasing carbon dioxide partial pressure reveals another positive engineering aspect of the working fluid. It operates to improve the efficiency of the oxygen unloading process in the systemic tissues where the blood takes on carbon dioxide produced by metabolic activity for transfer in the lungs. This improvement is easily seen by noticing that, for values of the oxygen partial pressure in the venous region, the shift accompanying the carbon dioxide uptake will lead to decreased oxygen saturation and, hence, increased unloading. The effect is not particularly pronounced for the resting state and average carbon dioxide uptake since, under these conditions, the partial pressure of the carbon dioxide increases from 40 mm Hg to only about 45 mm Hg. This corresponds to a shift of only about 20% of that indicated in Fig. 2.2. However, for greater uptakes from more active tissues or exercise, the effect clearly can become more significant.

As indicated earlier, shifts of the oxygen-dissociation curve under standard carbon dioxide conditions are also found when comparing curves from mammals of different size. The typical results shown in Fig. 2.3 demonstrate that the curves for mammals smaller than the human generally lie to the right of the standard curve for humans, while those for larger mammals generally lie to the left. This means that, for a given level of saturation, the partial pressure of the oxygen will be greater for smaller mammals than for large ones. Since, as noted in Chapter 1, smaller mammals have higher rates of oxygen consumption per gram of average tissue weight than larger ones, we thus see that the physiological characteristics reflected in the oxygen-dissociation curves are adjusted to assist in this requirement through increased unloading pressure.

Carbon Dioxide Transport and Exchange

When the blood passes through the systemic tissues, it not only unloads oxygen but picks up excess carbon dioxide for subsequent discharge in the lungs. Under resting conditions, about 4 ml of carbon dioxide is taken up and transported to the lungs for discharge in each 100 ml of blood. As noted above, this corresponds to an average increase of about 5 mm Hg in the partial pressure of the carbon dioxide from its arterial value of 40 mm Hg. The carbon dioxide is carried by the blood in three forms: (1) in the dissolved state; (2) in reversible combinations with the hemoglobin and the plasma proteins; and (3) in reversible bicarbonate form. The first two account for less than one-third of the carbon dioxide trasported, with the remainder being carried in the bicarbonate form.

The forms of carbon dioxide uptake in a systemic capillary are illustrated in Fig. 2.4. The carbon dioxide released by cell metabolism rapidly diffuses across the capillary wall, where a portion enters the plasma part of elemental volumes of blood as dissolved gas and a portion combined with the plasma proteins. The remainder diffuses into the red blood cells where a part combines with the hemoglobin and a part combines with water to form carbonic acid, which dissociates into bicarbonate and hydrogen. Much of the bicarbonate subsequently diffuses out of the cell into the plasma. On reaching the lungs, reverse processes occur, allowing the blood to discharge excess carbon dioxide.

Although the details of the transport process for carbon dioxide differ from those for oxygen, the overall uptake and discharge of carbon dioxide by diffusion into and out of the blood is similar. The average rate of diffusion across the capillary walls is thus described by a diffusion relation like that of Eq. (2.1) for oxygen transfer, and scaling relations will be similar to those established from this equation.

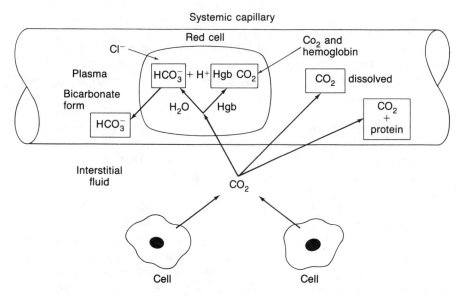

Fig. 2.4 Illustration of forms of carbon dioxide transport by the blood. The arrows indicate direction of the processes with systemic blood. Reverse conditions occur with pulmonary blood.

Exchange of Water and Dissolved Products

While the exchange of oxygen and carbon dioxide between the blood and its surroundings is the result of diffusion processes occurring directly through the capillary walls, water and many dissolved substances are insoluble in these walls and, hence, must move into and out of the blood by different means. Such means are provided by small water-filled pores penetrating the capillary walls. Small molecules are thus readily transported back and forth through these pores by diffusion. Water and dissolved substances are also filtered out of the blood and reabsorbed by mechanical flow through the pores.

Diffusion of molecules through the pores is governed by Fick's Law (see Appendix D), which relates the rate of mass transfer to the difference in concentration of the substance (expressed as mass per unit volume) and the area and distance over which the diffusion occurs. Thus, if ΔC denotes the averaged difference along the capillary between concentrations inside and outside the capillary, the average mass transferred per unit time by diffusion, say J_S, is described by the equation

$$J_S = D_S \frac{A_p}{A_c} \frac{\Delta C}{h_c} A_c \tag{2.3}$$

where, as earlier, A_c denotes the lateral surface area of the capillary and h_c denotes its wall thickness. Here also, A_p denotes the net cross-sectional flow area of the pores and D_S denotes a constant. Under change of scale, the ratio A_p/A_c may be assumed to remain constant in the capillary design, in which case the scaling relations derived from this equation will be analogous to those discussed earlier in connection with Eq. (2.1).

In the case of filtration and reabsorption by flow through the pores, there are two main driving forces involved. One arises from the blood pressure in the capillary relative to the fluid pressure in the surrounding tissue. The other arises from the plasma proteins which are too large to pass through the capillary pores and, hence, develope a natural tendency to attract water into the capillaries by the chemical process of osmosis. The pressure developed by the proteins is referred to as the colloid osmotic pressure.

At the arterial end of a typical systemic capillary of the human, the net blood pressure is about 37 mm Hg relative to the hydrostatic tissue pressure, while the net colloid osmotic pressure is about 28 mm Hg. Thus, at this end there is a net driving pressure of about 9 mm Hg tending to cause the water and dissolved substances to filter out of the capillary. In contrast, at the venous end of the capillary, the net blood pressure is about 20 mm Hg and the net colloid osmotic pressure remains at the value of 28 mm Hg. At the venous end, the net driving pressure of 8 mm Hg therefore acts inward and causes water and dissolved substances to be reabsorbed into the blood. In this way, important exchanges between the blood and its surroundings can occur while maintaining the water content of the blood relatively constant.

The process is illustrated in Fig. 2.5. It is normally assumed, consistent with the pressure values indicated, that outflow from the capillaries exceeds inflow by a small amount and that the resulting excessive fluid in the tissues is returned to the blood through the special lymph ducts that ultimately drain into the venous system.

Interestingly, the filtration and reabsorption of fluid in the kidneys differ from that just described for a typical systemic capillary in that filtration occurs in one set of capillaries and reabsorption in another connecting set. This is achieved by encasing the filtration capillaries in small capsules, with about 50 per capsule. These capsules collect the filtrate and allow it to drain through tubules which are surrounded by reabsorbing capillaries. Most of the filtrate from the first set of capillaries, together with secreted substances, is then reabsorbed by the second set. The remainder forms urine that drains onward to the bladder for discharge.

For an engineering description of the mechanical flow through the pores, we let ΔP denote the average driving pressure in a capillary. The average flow J_f of fluid through the pores is then expressible from fluid mechanics (Pappenheimer et al.1951) as

$$J_f = \frac{\beta \pi r_p^4}{8 \mu_f} \frac{\Delta P \, A_c}{h_c} \tag{2.4}$$

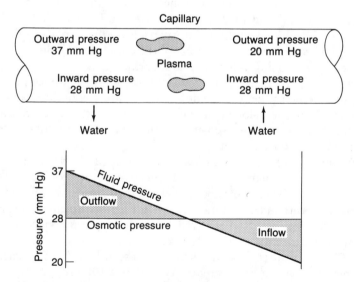

Fig. 2.5 Illustration of exchange of capillary water and dissolved substances with surroundings. Data source: Guyton (1971).

where, as before, A_c denotes the lateral surface area of the capillary and h_c, its wall thickness. Also, r_p denotes the radius of the pore cross-section, μ_f denotes the viscosity coefficient of the fluid moving through the pores, and β denotes the number of pores per unit of capillary area. Under change of scale, the first ratio on the right-hand side of this equation can be expected to remain fixed, so that scaling relations from this equation are again analogous to those from Eq. (2.1).

Mechanical Properties

It will be instructive to examine some additional properties of our working fluid, namely the mass density and viscosity of an elemental volume of blood. In contrast to the properties described above, these may be characterized as mechanical in nature. They are directly related to the inertial and frictional resistance of the blood to flow.

The mass density of the blood refers to its average mass per unit volume. Simple measurements indicate a constant value of 1.05 g/cm³ for all mammals, regardless of size. Thus, the density of blood is only slightly greater than that of pure water at room temperature (about 1 g/cm³).

The viscosity of the blood is more difficult to characterize than its density. As noted in Chapter 1, it is measured by the viscosity coefficient relating resistive force to flow for a given fluid. More specifically, it describes the resistance to shearing, or sliding, motion of the blood (see Appendix D). It is of major importance in describing the frictional-like resistance to blood flow in the vascular system. If, as in Fig. 2.6, we consider a sample of blood confined between a lower fixed plate and an upper moving plate, we may think of layers of the blood sliding over one another with linearly increasing velocity. The shearing stress τ acting on the blood is defined as the ratio of the tangential force F on the upper plate to the area A over which it acts, and the shearing rate $\dot{\gamma}$ is defined as the ratio V_o/d where V_o is the velocity of the upper plate and d is the distance between the two plates. The viscosity coefficient μ (or for brevity, the viscosity) is then defined by the ratio $\tau / \dot{\gamma}$.

For simple substances like water, the viscosity so defined is found to be independent of shearing rate and, hence, constant for a fixed testing environment. In the case of blood, however, the viscosity is found to vary with shearing rate in the manner shown in Fig. 2.6. Because of this variation, the viscosity is often referred to as the apparent viscosity at any given shearing rate. Interestingly, it can be seen from the viscosity variation in Fig. 2.6 that the blood viscosity does, in fact, become essentially constant for shearing rates greater than about 100 sec⁻¹, its value being about 0.032 dynes-sec/cm². As will be shown later, this is indeed the case for blood flow in the smaller vessels of mammals where viscous effects are

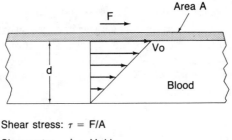

Shear stress: $\tau = F/A$

Shear rate: $\dot{\gamma} = V_0/d$

Viscosity: $\mu = \tau/\dot{\gamma}$

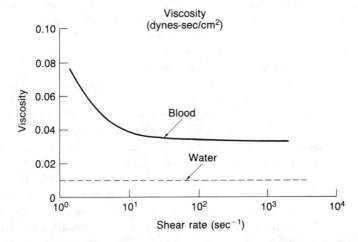

Fig. 2.6 Typical variation of the viscosity of blood with shear rate. Data source: Cokelet (1972).

of utmost importance. For this case, the expression between pressure difference and flow in a vessel is the famous Poiseuille equation described in Appendix D.

In connection with this last remark, it is of interest to note that experiments with steady blood flow through tubes of very small diameter have indicated that the average value of the blood viscosity does, in fact, decrease with decreasing tube diameter even though the shearing rates are sufficiently high to preclude rate effects of the kind indicated in Fig. 2.6. The effect is referred to as the Fahraeus-Lindquist effect after its discoverers. It has now been attributed to a reduced concentration of red blood cells in the smaller diameter vessels as a result of entrance conditions from the feed reservoir (Cokelet 1972). With fewer red blood cells present, the resistance to shearing motion is reduced and the viscosity of the blood is decreased toward the smaller value existing for the plasma alone, about 0.015 dynes-sec/cm^2. However, insofar as the actual non-steady blood flow in the smaller vessels of the circulatory system is concerned, the magnitude of the decrease, if any, is unknown. The effect will therefore be neglected in the present work and the viscous characteristics of blood flow in small vessels will be assumed to be approximately the same as those in larger ones.

2.3 THE PUMPS—THE HEART

From our earlier discussions, we know that the heart consists, in engineering terms, of two pumps acting in series in the cardiovascular system in order to circulate the blood through the systemic and pulmonary vascular beds. A sectional view of the heart is shown in Fig. 2.7. As indicated in this figure, each side of the heart consists of two chambers, the atrium (or auricle) and the ventricle. The atrium serves as a reservoir for incoming blood, while the ventricle provides the main pumping action. Valves between the ventricle and connecting artery (semilunar valve) and between the atrium and ventricle (A-V valve) control the direction of blood flow and maintain a positive pressure in the arteries and connecting vessels.

The walls of the heart chambers are composed of specialized muscle referred to as cardiac muscle. This muscle, like skeletal muscle, consists of striated fibers with diameters typically of the order of 0.01 mm in the human heart. The individual fibers respond to electrical impulse by shortening, causing in turn a contraction of the overall muscle. Unlike skeletal muscle, however, where the fibers are arranged in parallel bundles, the cardiac muscle fibers form a complex mesh network with numerous end connections. This allows stimulation at one point to spread rapidly to other parts of the muscle.

An illustration of the fibers in cardiac muscle is shown in Fig. 2.8. Electron microscopy studies have revealed that the fibers consist, in fact, of series-connected cardiac cells. As indicated in Fig. 2.8, these cells contain small bundles of rods, called fibrils, which contract under excitation, then causing in turn the overall fibers to contract.

Pumping action of the heart arises from cyclic contractions and expansions of the left and right venticles. The contractions and expansions of the two ventricles are synchronized so that blood is squeezed into the lungs by the right ventricle at the same time that blood is squeezed into the other parts of the body by the left ventricle. Blood circulation through the lungs is referred to as pulmonary circulation while that through the other parts of the body is referred to as systemic or peripheral circulation.

During contraction, the ventricle-artery valves are forced open and the atrium-ventricle valves are forced shut, as a result of their design, so that blood can be pushed through the system. Near the end of the contrac-

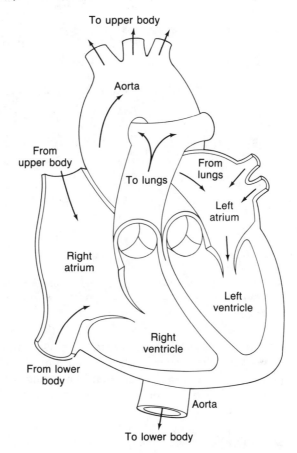

Fig. 2.7 Illustration of sectional view of the heart, with arrows indicating the direction of blood flow.

tion, the pressure in the ventricle begins to fall, and the ventricle-artery valve closes with the onset of backward flow. With subsequent expansion of the ventricle, the atrium-ventricle valve opens, allowing blood to flow into the ventricle in preparation for the next cycle of pumping. The periods of contraction and expansion of the ventricles, referred to, respectively, as systole and diastole, are approximately equal. More precisely, contraction involves about 40% of the cardiac cycle and expansion involves the remaining 60%.

Figure 2.9 illustrates the pressure, atrium- and ventricle-volume variations that occur in the left heart of man during a normal cardiac cycle. As indicated, there is a small interval at the beginning of systole where the pressure rises without change in ventricular volume. This is known as the isometric-contraction period of the cardiac muscle. The ten-

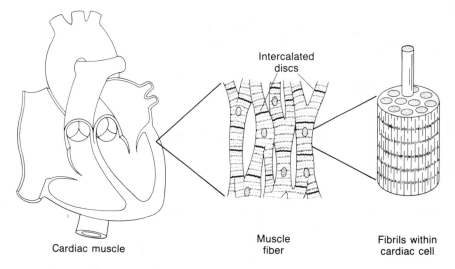

Intercalated
discs

Cardiac muscle   Muscle   Fibrils within
                                 fiber   cardiac cell

Fig. 2.8   Illustration of the structures of cardiac muscle fiber and the cardiac cell. The intercalated discs shown are the end membranes separating cardiac cells.

sion in the muscle increases without change in the muscle dimension. Figure 2.9 also shows that excitation of the cardiac muscle begins in the atrium and rapidly propagates to the ventricle. This causes an atrial contraction which forces additional blood into the ventricle to complete the filling process prior to the start of the next cycle of pumping.

Cardiac Excitation

The periodic contractions of the heart have their origin in the heart itself rather than in nerve stimuli from the brain or spinal cord, as might at first be thought. The rhythmic pumping of the heart of a mammal can, in fact, be maintained for hours after removal from the body, provided only that blood, or a warm oxygenated solution, is pumped through its vascular system. Although not essential for the rhythmic contractions of the heart, nervous regulation does exist for the purpose of fine tuning to meet the body's needs. This control is provided by the autonomic nervous system which supplies nerves from the brain and various parts of the spinal cord to the heart, and other major organs of the body, for involuntary control. The sympathetic nerves of this system act to increase heart rate and cardiac output in response to the body's needs, while the parasympathetic nerves provide opposite control by acting to reduce heart rate.

The origin of the basic periodic contractions of the heart lies in a small region of the upper right atrium, known as the S-A node. Within this region, periodic electrical discharges, or action potentials, are generated

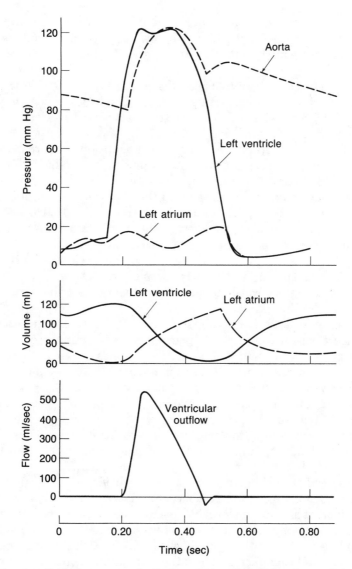

Fig. 2.9 Typical pressure and volume variations in the left atrium and ventricle of the human. Outflow from the left ventricle is also illustrated. Data source: Selkurt and Bullard (1971).

by specialized muscle fibers which ultimately initiate the rhythmic cardiac contractions. Each cardiac discharge, generated in the S-A node, rapidly spreads to the atria, causing contraction of these chambers. The signal is then delayed somewhat at a second node, the A-V node, before being conducted into the ventricles. This delay allows time for the atria to contract and complete the filling of the ventricles before their contraction starts. After the delay, the signal enters the ventricles and quickly spreads across them through the so-called Purkinje fibers. The conduction of the signal through these fibers is approximately four times faster than it would be through ordinary cardiac muscle fibers, so that the presence of the Purkinje network ensures a nearly simultaneous contraction of all parts of the ventricles in response to the electrical discharge.

The physical basis for the rhythmic electrical discharges from the S-A node is not well established. They are thought to be the result of the selective movement of sodium and potassium ions into and out of the cells comprising the fibers of this region. Such movement can explain the periodic electrical discharge and recharge of the fibers such as observed with potential measurements of the kind illustrated in Fig. 2.10. However, the means by which the ions are transported back and forth across the surface membrane of the cells are not understood. The movement of positive ions into the fiber cells is, of course, consistent with the electrical attractive force of the negative interior that is revealed by the potential measurements. But the net outward movement to recharge the fiber requires an energy source of some kind.

The electrical discharges illustrated in Fig. 2.10 are those for the human heart with resting rate of about 70 beats per minute, corresponding to a time between beats of about 0.9 sec. This heart rate is the result of some restraint imposed on the basic ion movements by an inhibitory substance (acetylcholine) released by the parasympathetic nerves. In contrast, during an emergency or exercise, the ion movement can be enhanced by release of an excitatory substance (norepinephrine) by the sympathetic nerves and this can, with strenuous exercise, cause an increase in heart rate by a factor of two or more. This translates to a decrease in the time between impulses by this same factor.

If, instead of the human heart, we now consider a small animal such as a mouse, we may recall from the measurements given in Chapter 1 that its resting heart rate is considerably greater than that of the human. For the mouse, for example, it is about 600 beats per minute. Presumably, exercise and nerve stimulation will likewise increase this by a factor of two or so. But what design factor caused the increase in the resting heart rate by a factor of almost nine over that of the human heart? The answer would appear to be in the design, or sizing, of the individual cells forming the muscle fibers. As indicated earlier with respect to general cardiac muscle, each fiber is known to consist of a single row of individual cells making end-to-end connection. If the size of each cell is reduced, its volume will also be reduced. It will, accordingly, take less ion movement into and out of the cells in order to generate the same ion change per cell volume and, hence, from electrochemical theory, the same potential change in the cardiac discharge. Assuming the same basic processes are operative, the

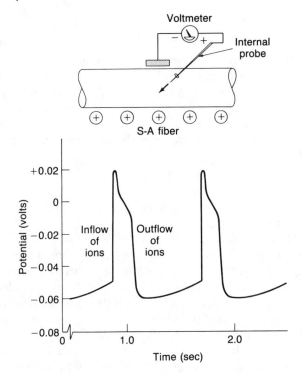

Fig. 2.10 Representative action-potential measurements from S-A fiber of human. Data source: Guyton (1971).

time between cardiac discharges will therefore be reduced and the rate of heartbeat increased.

We may put this idea on a quantitative basis by assuming a design description for the ion-transport process analogous to that for diffusion (see Appendix D). Thus, in fundamental terms, we consider a typical cell of the fiber and assume the rate of movement J_E of ions across its surface membrane to be described by the equation

$$J_E = D_E \Delta F_E \frac{A_m}{h_m} \qquad (2.5)$$

where A_m denotes the membrane area, h_m denotes its thickness, ΔF_E denotes the driving force, and D_E denotes a proportionality constant.

If we now let ℓ_s denote any typical length dimension of the cell, such as its radius or length, and assume all dimensions are changed by the same factor under change in mammal size, the membrane area will vary with ℓ_s^2, the thickness with ℓ_s, and the cell volume with ℓ_s^3. Using ω to denote the heart rate, we then have from Eq. (2.5) the proportional relation

$$\frac{J_E}{\omega \ell_s{}^3} \, \alpha \, \frac{\Delta F_E}{\omega \ell_s{}^2}$$

The ratio on the left-hand side denotes the ion movement per cell volume into or out of the cell during a cardiac cycle, which, as we noted earlier, must be constant under size change for constant potential change of the cardiac discharge. With the driving force likewise constant under size change, this proportional relation requires that the product $\omega \ell_s{}^2$ be constant. Thus, the heart rate must, under the assumptions made, follow the scaling relation

$$\omega \, \alpha \, \ell_s^{-2} \tag{2.6}$$

that is, the heart rate must vary inversely with the square of the cell's linear dimensions.

Returning to the design of the mouse, we now see that its increased resting heart rate over that of the human can be achieved by simply reducing the size of its S-A fiber cells by about one-third. From relation (2.6), this reduction is then predicted to yield a heart rate about nine times faster than the human rate, consistent with the measurements of Chapter 1. More generally, we shall find additional scaling restrictions in Chapter 3 that will require cardiac cell size of mammals to vary with body weight raised to the one-eighth power. The scaling relation (2.6) will then be found to predict heart rate to vary inversely with body weight raised to the one-fourth power, in agreement with the measurements and associated empirical scaling law described in Chapter 1.

Cardiac Muscle Mechanics

The cyclic contractions of the heart ventricles occur as a result of cyclic shortening of surrounding muscle. As noted earlier, such cardiac muscle is fibrous in nature and responds to stimulus by shortening along the longitudinal direction of the fibers. If, in an idealized sense, we think of the ventricles as cylindrical in form, with the muscle fiber running in the circumferential direction, we see that the muscle shortening will cause radial contraction of the walls of these chambers and give rise to the pumping action described earlier.

From a mechanical point of view, the muscle fibers are generally envisioned to consist of a contractile element in series with an elastic one. When the muscle is under load and subject to stimulus, the contractile element is first imagined to shorten, causing a progressive stretching of the series elastic spring until the force in it equals either the maximum force developable by the muscle or the total load on the muscle (preload plus afterload) if this is less. During this phase, the overall length of the muscle does not change and the phase is referred to as the isometric phase of contraction. If the maximum force is not achieved in this phase, further shortening of the contractile element under constant loading occurs and the muscle itself shortens. The latter stage is referred to as the afterload isotonic phase of muscle contraction.

Interestingly, experiments have shown that the amount of afterload that a muscle is able to carry increases with increasing preload, that is, with increasing initial stretch of the muscle fiber before contraction. This is the basis for the famous Starling Law of the Heart, proposed in 1915 by the physiologist Ernest Starling (1866–1927), which states that the energy of contraction of the heart is a function of the length of the muscle fibers before contraction.

The idea of contractile and elastic elements of muscle fiber is convenient for describing the muscle response to stimulus but does not lend itself to simple quantitative characterization. For this purpose, we develop here an alternate engineering description that allows the two elements to be combined into a single generalized expression.

We consider a strip of cardiac muscle and assume it acts mechanically as a non-linear elastic spring of natural (unstretched) length L_o, as shown in Fig. 2.11(a). For analytical purposes, we further assume the elastic response (force vs deflection) characteristics of this spring to be expressible as a simple quadratic relation. Such a function will, in fact, be seen later to provide a good representation of cardiac muscle response over an appreciable range of deflections. In the absence of stimulus, that is, in the passive state, we accordingly require that a preload force F_o applied to the strip will give rise to a corresponding change in length (or extension) ΔL_o such that the following equation is satisfied

$$F_o = k \left(\frac{\Delta L_o}{L_o} \right)^2 \tag{2.7}$$

where k denotes a constant for the strip. Under stimulus, that is, in the active state, we now suppose a physiological change of some kind to occur in the muscle such that its natural length is reduced to a new length, described by $L_o / \sqrt{1+S}$, where S is defined as the contractile factor. If we denote the additional afterload carried by the muscle by ΔF, the total load by $F_1 = F_o + \Delta F$, and the value of the contractile factor needed to carry this load by $S = S_1$. We then have from Eq. (2.7) the relation

$$F_1 = k \left(\frac{\Delta L_o}{L_o} \right)^2 (1 + S_1) \tag{2.8}$$

which applies to the isometric stage of loading where the overall muscle length does not change.

Now suppose further physiological change in the muscle occurs such that the contractile factor increases to its maximum value S_o. If there is no further increase in afterload, the length of the muscle must then de-

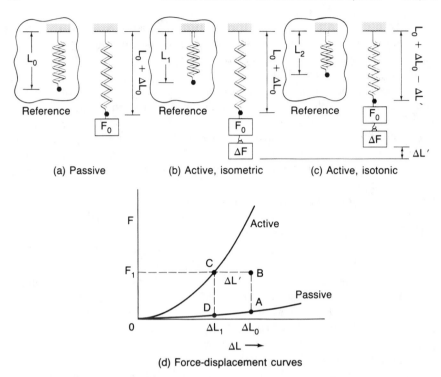

(a) Passive   (b) Active, isometric   (c) Active, isotonic

(d) Force-displacement curves

Fig. 2.11 Mechanical representation of cardiac muscle by a non-linear spring model.

crease. From Eq. (2.7), we thus have the relation for this isotonic phase of loading expressible as

$$F_1 = k \left(\frac{\Delta L_o}{L_o} - \frac{\Delta L'}{L_o} \right)^2 (1 + S_o) \qquad (2.9)$$

where $\Delta L'$ denotes the afterload shortening of the muscle.

The isometric and isotonic stages of loading described by Eqs. (2.8) and (2.9) are illustrated in Fig. 2.11(b) and (c). In the passive state, the free length is denoted by L_o, the load by F_o, and the preload extension by ΔL_o. In the active isometric stage of loading, the free (reference) length is reduced to a value given by $L_1 = L_o/\sqrt{1 + S_1}$, the extension is still of the value ΔL_o, and the load is increased by the amount ΔF to the new value given by $F_1 = F_o + \Delta F$. Finally, in the isotonic stage, the free length is reduced to the value given by $L_2 = L_o/\sqrt{1 + S_o}$, the extension is reduced to the value $\Delta L_o - \Delta L'$, and the load is held fixed at the value F_1.

We assume the total contractile factor S_o to be constant, independent of preload, in which case we see that Eq. (2.9) requires the final force developed by the muscle to depend only on its final extension, $\Delta L = \Delta L_o - \Delta L'$, and to be independent of the path of loading. Thus, in Fig. 2.11(d), the point C, corresponding to load F_1 and extension $\Delta L_1 = \Delta L_o - \Delta L'$, may be reached by the isometric contraction from A to B, followed by the isotonic contraction from B to C; by isometric contraction from D only; or by any other chosen path. The relation given by Eq. (2.9) with ΔL substituted for $\Delta L_o - \Delta L'$ therefore represents an equation of state of the muscle. Such a path-independent condition does, in fact, exit for cardiac muscle (Downing and Sonnenblick 1964). Interestingly, it is, however, not generally the case for skeletal muscle (Sagawa 1973).

We may illustrate the validity of this path-independent condition for cardiac muscle using the original experimental measurements of Downing and Sonnenblick (1964). These are shown in Fig. 2.12, where values of the developed force in both the passive and active state are plotted against the stretched length of a cardiac muscle segment. The data in the active state are from both pure isometric and pure isotonic response as well as from afterload isotonic response. For example, the data point I in the active state resulted from pure isometric response from the data point C of the passive state. In contrast, the data point E of the active state resulted from pure isotonic response from the same data point C in the passive state. Also, as indicated, the data point F in the active state resulted from isometric response from points C to D, followed by afterload isotonic response to point F. We see that the various data points in the active state all fall on the same curve, consistent with our previous discussion of the path-independent condition of Eq. (2.9).

In addition to demonstrating the path-independence of cardiac muscle response, we may illustrate with these same measurements the general validity of the simple quadratic force-deflection response assumed in the development of Eqs. (2.7) through (2.9). For this purpose, we first determine, for each of the force values, the change in length ΔL of the muscle strip from its effective initial value L_o, estimated as 9.65 mm. We may then plot values of the measured force for the active and passive states against the corresponding values of the square of the ratio $\Delta L/L_o$. If the quadratic form applies, the resulting plots of the data points should fall on straight lines defining the active and passive responses.

Such plots are shown in Fig. 2.13, where we see that the data points do indeed fall closely around straight lines defining the two responses. These results thus confirm the general acceptability of the simple quadratic forms of Eqs. (2.7) through (2.9) as good engineering approximations. It is interesting to notice that the slopes of the lines representing the active and passive states are about 90 and 5, respectively. These values accordingly provide a value of the term $1 + S_o$ in Eq. (2.9) of 18. In

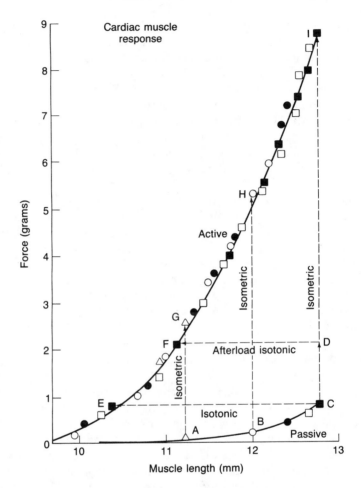

Fig 2.12 Force-length measurements for passive (lower) and active (upper) cardiac muscle response, illustrating equation-of-state nature of the active state. Data source: Downing and Sonnenblick (1964).

terms of the present formulation, this means that the natural length of the muscle is reduced under stimulus by a factor of $1/\sqrt{18}$, that is, by a factor of about one-fourth.

The above considerations have dealt with the force-displacement response of cardiac muscle and its appropriate engineering description. In studying muscle-response characteristics, it is also customary to consider the initial velocity of shortening of the muscle under varying amounts of afterload. It is of interest to include this response behavior in the present description.

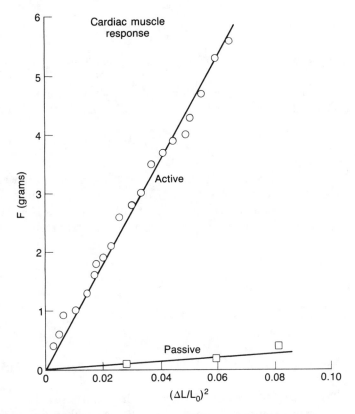

Fig. 2.13 Measured force data of Downing and Sonnenblick (1964) plotted against the square of the fiber strain $\Delta L/L_o$, showing applicability of a simple quadratic representation

To do this, we first modify Eq. (2.9) so that it gives us information on the developing stages of the muscle shortening $\Delta L'$. In particular, we replace the maximum contractile factor S_o in Eq. (2.9) with the intermediate sum $S_1 + S_2$. We regard S_1 as fixed in order to carry the constant load F_1 and S_2 as variable, with its value increasing with time from zero to its final maximum value associated with the maximum value S_o used in Eq. (2.9) in our earlier discussion. In this way, the muscle shortening $\Delta L'$ must also be considered variable, with its value increasing with time from zero to its final maximum value. Forming rates (time derivatives) from this equation, we can then find, with the help of Eqs. (2.7) and (2.8), that the initial velocity V of muscle shortening (or initial time rate of change of shortening) is expressible, for the initial conditions $\Delta L' = 0, S_2 = 0$, as

$$V = \frac{F_o \Delta L_o \dot{S}_2}{2F_1} \tag{2.10}$$

where $\dot{S}_2$ denotes the initial rate of change of the contractile factor with time.

We may place this last equation in a more conventional form by expressing the contractile factor $\dot{S}_2$ as

$$\dot{S}_2 = \frac{2F_1}{F_o \Delta L_o} \frac{V_c(F_m - F_1)}{F_1 + F_c} \tag{2.11}$$

where F_c and V_c denote parameters and F_m denotes the critical force value for pure isometric response, as determined from Eq. (2.9) when the muscle shortening $\Delta L'$ is set equal to zero. Substituting this relation into Eq. (2.10), we then find that the expression for the initial shortening velocity can be written as

$$V = \frac{V_c(F_m - F_{1)}}{F_1 + F_c} \tag{2.12}$$

or, alternatively, in standard form (Hill 1938) as

$$(F_1 + F_c)(V + V_c) = (F_m + F_c)V_c \tag{2.13}$$

which is known to apply to both skeletal and cardiac muscle (Sagawa 1973).

An illustration of the applicability of this relation is shown in Fig. 2.14 using measurements of Sonnenblick (1964) corresponding to two different preloads. The values of the parameters in Eq. (2.13) were estimated from the data and are as indicated in the figure. With these values, we see that the equation provides a good representation of the measurements for both preloads. It is interesting to notice that the maximum (extrapolated) velocity of shortening, corresponding to zero load on the muscle, is independent of the preload. This is a general result noted by Sonnenblick (1962, 1964). In terms of Eq. (2.12) or (2.13), we have the relation

$$V_{max} = \frac{V_c F_m}{F_c} \tag{2.14}$$

so that the ratio on the right-hand side is also independent of preload.

Before concluding our discussion of cardiac muscle response, it is worthwhile to set down the general relation from the theory that describes the force in the ventricular walls of a beating heart when the afterload is not strictly constant. This relation follows from Eq. (2.7) in the same manner as Eq. (2.9), but with a general force F rather than the constant force F_1 and, as above, a variable afterload contractile factor S_2. Thus, we have the equation

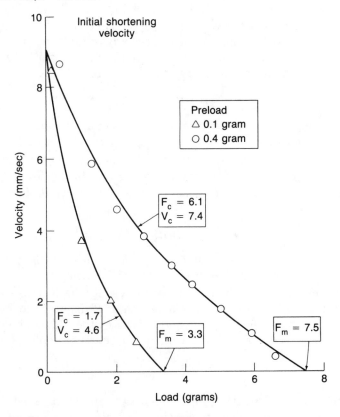

Fig. 2.14 Typical measurements of the initial shortening velocity of heart muscle by Sonnenblick (1964) for two different preloads and various afterloads. The parameters F_c, V_c are estimates based on the for of Eq. (2.13).

$$F = k\left[\frac{\Delta L_o}{L_o} - \frac{\Delta L'}{L_o}\right]^2(1 + S_1 + S_2) \qquad (2.15)$$

which may be written with the help of Eq. (2.7) and the definition $S'_o = S_1 + S_2$ as

$$F = F_o(1 + S_1) + F_oS_2 - k(1 + S'_o)\left(2\frac{\Delta L_o}{L_o} - \frac{\Delta L'}{L_o}\right)\frac{\Delta L'}{L_o} \qquad (2.16)$$

Now, the first term on the right-hand side of this equation is seen from Eq. (2.8) to be the force associated with isometric contraction. The second term is the additional force caused by further contraction of the natural (reference) length of the muscle and the third term is the reduction in force resulting from change in extension of the muscle,. Over a pumping cycle of the heart, the isometric force $F_o(1 + S_1)$ may be written simply as F_s. If ω denotes the frequency of heartbeat (muscle contraction) and t denotes the time, we may also write the variable force $F_o S_2$ in terms of a general function $G(\omega t)$ as $F_o G(\omega t)$. Finally, the last term may be written in terms of the general function $H(\varepsilon)$, where ε represents the variable ratio $\Delta L'/L_o$, as $k(1 + S_o')\, H(\varepsilon)$. The equation is then seen to take the form

$$F = F_s + F_o G(\omega t) - k(1 + S_o')H(\varepsilon) \qquad (2.17)$$

In this equation, we may, for a simplifying approximation, replace the variable contractile factor $S_o' = S_1 + S_2$ with its average value, say S_o'', over a pumping cycle. If we further assume an idealized cylindrical form for the ventricle, with muscle fibers running in the circumferential direction, we may interpret this last equation on a per-unit-length basis and write the equation as

$$F = F_s + F_o G(\omega t) - Eh\, H(\varepsilon) \qquad (2.18)$$

where E denotes an elastic modulus of the muscle and h the ventricular wall thickness such that $Eh = k(1 + S_o'')/\ell$, with ℓ being the length of the ventricle. Here the ratio ε, or strain, now represents the decrease in length of the circumference per unit of initial length. If U denotes the inward radial displacement of the ventricle and a denotes the radius about which the displacement occurs, we have, from geometry, the simple relation

$$\varepsilon = \frac{\Delta L'}{L_o} = \frac{2\pi a - 2\pi(a - U)}{2\pi a} = \frac{U}{a} \qquad (2.19)$$

Thus, the ratio ε is directly proportional to the inward radial displacement. We shall have occasion to use these last equations later in Chapter 3, when we consider an engineering design model for the cardiovascular system.

2.4 THE PIPING NETWORK—THE VASCULAR BEDS

We have previously characterized the systemic and pulmonary vascular beds simply as a series of branching arteries conducting blood from the heart to the capillaries and a series of coalescing veins draining blood from the capillaries back to the heart. While this picture gives a general view of the geometry of the vascular beds serviced by the left and right sides of the hearts, a more detailed description considers blood flow to the various organs of the body and divides the arteries and veins into several discrete categories based on vessel diameter.

Figure 2.15 illustrates the general path of blood flow in the systemic and pulmonary systems when account is taken of the main organs and organ systems. As seen here, and noted in earlier discussions, a small amount of the flow out of the left ventricle is diverted to the non-aerated

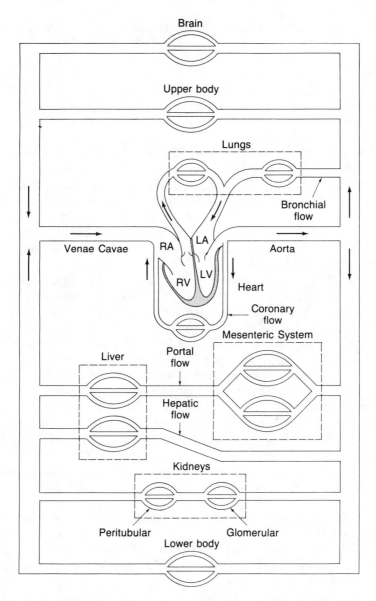

Fig. 2.15 Illustration of the cardiovascular system showing details of blood flow to and from various organs.

capillaries in the lungs through the bronchial flow and returned directly to the left ventricle where mixing (venous admixture) and further circulation through the systemic system occurs. Coronary flow supplying the cardiac muscles with oxygen is also shown. In the case of blood flow to liver, we see that a portion consists of venous blood draining from the mesenteric system (the stomach, intestines, pancreas, and spleen) in the portal flow. The remaining blood supply to the liver consists of arterial blood entering through the hepatic flow.

The flow of blood through each of the two kidneys is illustrated in Fig. 2.15 by a single representative network. The flow is seen to be directed first through the glomerular capillary system, where filtration of fluids occurs. It is then directed through the series-connected peritubular capillary system where valuable substances are reabsorbed. Flows depicted in the figure to the upper and lower body include blood supply to muscle and other organs. Circulation to the brain is, however, singled out to emphasize the importance of blood supply to this organ.

Blood flow to the capillary beds of the various organs, although shown in Fig. 2.15 as being through single connecting vessels, is, in fact, through a series of branching vessels. Return flow to the heart is likewise through a series of coalescing vessels. Thus, the main connecting artery (the aorta or pulmonary artery) from a ventricle can be regarded as first branching into a number of large arteries in order to direct the blood flow to the various organs. The large arteries may next be considered to branch into thousands of small arteries that subsequently branch into many more very small arteries, called arterioles. These vessels feed the vast number of capillaries branching from them. On the venous side, the capillaries first coalesce into very small veins, called venules, which, in turn, may be considered to coalesce into the small veins. The small veins are next considered to converge into the large veins, which finally join the great veins (venae cavae and pulmonary veins) that connect directly to the heart.

The above description is, of course, highly idealized. With the exception of the main connecting vessels, there is no sharp transition from one class of arteries or veins to the next. Rather, vessels branch or coalesce with only small variations in diameter. In the spirit of the present description, we may, however, assume any vessel to give off or receive vessels of the next smaller category at various locations along its length.

An illustration of this idealized description for the systemic vascular system is given in Fig. 2.16. The aorta is shown branching into a number of large arteries, each of which directs blood to a major systemic location. A typical large artery is then shown, with its branchings into small arteries. These arteries further direct blood flow to the tissues of a general systemic location. The branching of a typical small artery into several

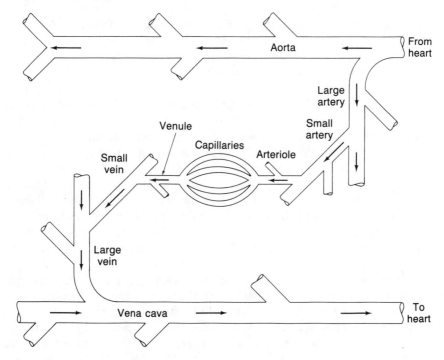

Fig. 2.16 Illustration of branching and coalescing vessels of the systemic vascular system.

arterioles is next shown, with one of these supplying blood to a representative capillary network where exchanges take place between capillary blood and surrounding tissue. The collecting vessels are also indicated in the figure. Venules from the capillaries are seen to merge into small veins which, in turn, merge into the large veins. These finally join one of the venae cavae for return of the blood to the heart.

Although not shown in Fig. 2.16, there also exist small constricting vessels at the entrance to the capillaries. These so-called precapillary sphincters act as valves to control the opening and flow of blood through the individual capillaries. In the resting state, they function to keep reserve capillaries closed and inoperative, while, in exercise, they allow them to be open and operative in order to meet increased demands of the tissues.

The dimensions of arterioles, capillaries, and venules are such that they cannot be seen without the assistance of a light microscope. These vessels are therefore referred to broadly as microscale vessels, in contrast with the other arteries and veins, which are referred to as macroscale vessels. Blood flow in the capillary networks is likewise referred to as microcirculation. In addition to the arterioles, capillaries, venules, and precapillary sphincters, studies have revealed other secondary vessels in the microcirculatory unit that aid in blood distribution and flow (see, for example, Friedman 1971).

Typical values of the dimensions and number of the various main vessels in the systemic vascular bed of humans are listed in Table 2.2. Guidance for these values is provided by extensive measurements of the mesenteric vasculature of the dog and their extrapolation to its total systemic bed (see, for example, Noordergraaf 1978).

It can be seen that each arterial branching produces a large number of additional vessels of reduced size. On the venous side of the capillaries, the reverse is true and the vessels converge to fewer and fewer ones of greater and greater dimensions until the venae cavae are reached, with radius somewhat greater than that of the aorta and total length about the same as the aorta length.

In spite of the decreasing size of the branching vessels on the arterial side of the capillaries, the increasing number of vessels can, in general, be seen to cause the total cross-sectional (flow) area of each group to increase. From the values listed in Table 2.2, we see, for example, that the sectional area of the aorta is about 4.5 cm^2, the net area of the small arteries is about 60 cm^2, and that of the capillaries is about 2400 cm^2. The increase in net area for each group of branching vessels is important because it causes the mean blood velocity to decrease progressively until at the capillaries it is only a small fraction of that in the aorta, thereby allowing time for exchanges in these vessels by diffusion and filtration.

TABLE 2.2   Estimated Geometry of the Systemic Blood Vessels in Humans

Vessel	Typical radius (mm)	Typical length (mm)	Typical number
Aorta	12	500	1
Large arteries	3	250	50
Small arteries	1	50	2×10^3
Arterioles	0.010	2	1×10^8
Capillaries	0.005	1	3×10^9
Venules	0.015	2	1×10^8
Small veins	2	50	2×10^3
Large veins	5	250	50
Venae cavae	15	250	2

Velocity Estimates

Some illustrative calculations of the entrance velocity of blood in the various vessels are of interest. The mean velocity of the blood entering a single vessel or a group of vessels may be estimated by dividing the mean flow rate by the net sectional flow area as illustrated in Fig. 2.17. Assuming a typical resting flow rate in the aorta of the human to be about 5500 ml/min., we thus find the typical blood velocities tabulated in Table 2.3.

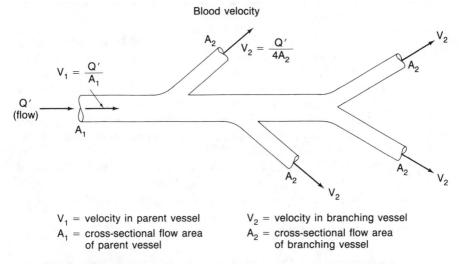

$$V_2 = \frac{Q'}{4A_2}$$

$$V_1 = \frac{Q'}{A_1}$$

V_1 = velocity in parent vessel V_2 = velocity in branching vessel
A_1 = cross-sectional flow area A_2 = cross-sectional flow area
        of parent vessel   of branching vessel

Fig. 2.17   Illustration of blood-velocity calculations in branching vessels.

From these calculations, we see that the average velocity of the blood passing through a capillary is about 40 times less than that in a small artery and about 500 times less than that in the aorta. This reduced velocity of 0.04 cm/sec provides, of course, an increase in the time over that otherwise available for an elemental volume of blood to be in a capillary vessel and subject to transfer processes. From Table 2.2, we see that the length of a capillary is about 0.1 cm, so the time that a particle of blood remains in it is approximately $0.1/0.04 = 2.5$ sec, in contrast with the much smaller value of 0.005 sec that would exist if the net area of the capillaries equaled that of the aorta.

Apparent Viscosity

We noted earlier that the viscosity of the blood is dependent on the shearing rate of the blood flow for small rates. We may use the velocity

TABLE 2.3.   Mean Blood Velocities in Various Systemic Vessels of Humans

Vessel	Area (cm^2)	Mean velocity (cm/sec)
Aorta	4.5	20
Small arteries	60	1.5
Capillaries	2400	0.04
Small veins	250	0.4
Venae cavae	14	7

estimates in Table 2.3 to examine the significance of this, if any, to the frictional resistance experienced by the blood in its circulation. For viscous flow in a straight tube of radius r, the frictional resistance is dominant at the tube walls. The associated shearing rate is determined by the ratio 4 V/r, where V denotes the mean velocity. Using the data in Tables 2.2 and 2.3, we thus find the values tabulated below in Table 2.4. Also shown are corresponding estimates of the apparent blood viscosity as determined from Fig. 2.6 given earlier.

It can be seen that the viscosity is essentially constant at the vessel walls for all vessels except the small veins and vena cava, where it is only about 20% greater than the nominal value of 0.032 dynes-sec/cm^2.

Resistance to Blood Flow

Let us now examine the resistance that the various vessels present to the blood flow. Of particular interest is the total resistance that must be overcome by ventricular pumping in order to move the blood through the arteries, capillaries, and veins to the collecting atrium reservoir. There are two main types of resistances to the blood flow: viscous and inertial.

TABLE 2.4.   Shear Rates and Apparent Viscosity of Blood in Various Vessels of the Human

Vessel	Shear rate (sec^{-1})	Viscosity (dynes-sec/cm^2)
Aorta	68	0.032
Small arteries	60	0.032
Capillaries	320	0.032
Small veins	8	0.039
Venae cavae	19	0.038

The viscous resistance is essentially the frictional resistance opposing the flow of the blood and the inertial resistance is that opposing the acceleration of the blood.

We consider first the viscous resistance. For a non-branching vessel of length L and radius r, this resistance is measured in terms of the pressure drop, say f_v, over its length, and is expressible from the famous Poiseuille Law (see, for example, Daugherty and Franzini 1965; also Appendix D) as

$$f_v = \frac{8\mu L Q'}{\pi r^4} \qquad (2.20)$$

where Q' denotes the flow (volume per second) in the tube and μ denotes the viscosity. In the case of a branching vessel, we may, as an engineering approximation, assume a constant radius and establish a formula for the pressure in a vessel with uniform branchings along its length, as illustrated earlier in Fig. 2.17. If Q' now denotes the entrance flow, the average flow along the branching tube will then be approximately $Q'/2$, and the net pressure drop will be simply one-half that given by Eq. (2.20). This estimate applies, of course, to a single vessel. However, for a class of vessels such as the large arteries, the pressure drop across each may be assumed about the same, so that Eq. (2.20) with Q' replaced by $Q'/2$ provides the pressure drop from viscous resistance across the entire class of vessels. Moreover, if Q denotes the total blood flow out of the heart and n_i denotes the number of vessels in a specified class, we have the flow Q' determined by the ratio Q/n_i. Thus, using Eq. (2.20) as a guide, we may express the pressure drop across any class of branching arteries as

$$f_{iv} = \frac{4\mu L_i}{\pi r_i^4} \frac{Q}{n_i} \qquad (2.21)$$

where the subscript i on f, L, r, and n refers to the values for each class, with $i = 1$ referring, say, to the aorta, $i = 2$ to the large arteries, $i = 3$ to the small arteries, and $i = 4$ to the arterioles.

The branching of the arterioles into the capillaries marks the end of the branchings in the vascular system. The capillaries themselves may be regarded as a parallel, non-branching bed, with pressure drop f_{5v} determined from Eq. (2.20) as

$$f_{5v} = \frac{8\mu L_5}{\pi r_5^4} \frac{Q}{n_5} \qquad (2.22)$$

where L_5, r_5 and n_5 denote, respectively, the length, radius, and total number of the capillary vessels.

After passing through the capillaries, the blood flows into the venous system for return to the heart. The merging of each set of smaller veins into the next large ones along their length can be regarded as the reverse of the branchings considered for the arteries, so that Eq. (2.21) will also apply to the venous system, with, say, $i = 6$ referring to the venules, $i = 7$ to the small veins, $i = 8$ to the large veins and $i = 9$ to the venae cavae.

Consider next the inertial resistance to blood flow in the various blood vessels. For a non-branching tube of length L and radius r, the net pressure f_m needed to overcome this resistance is expressible (see, for example, Daugherty and Franzini 1965; also Appendix D) as

$$f_m = \frac{\varrho L \dot{Q}'}{\pi r^2} \tag{2.23}$$

where ϱ denotes the mass density of the blood and $\dot{Q}'$ denotes the rate at which the flow in the tube is changing with time. This relation is based on the assumption that each elemental volume of blood in the tube is accelerated simultaneously. In fact, however, when the heart forces blood into the aorta, a pressure wave is generated which propagates down the vessel, leaving accelerated blood in its wake. Details of this dynamic aspect of blood flow have been reviewed by Noordergraaf (1978), among others. For the present work, we consider Eq. (2.23) to provide an approximate average measure of the inertial resistance.

For a branching tube, we may apply Eq. (2.23) in the same approximate manner as above for viscous resistance. We then find, analogous to Eq. (2.21), the relation

$$f_{im} = \frac{1}{2} \frac{\varrho L_i}{\pi r_i^2} \frac{\dot{Q}}{n_i} \tag{2.24}$$

where f_{im} denotes the inertial resistance in an artery or vein, $\dot{Q}$ denotes the rate of change of the total blood flow out of the heart, and the remaining symbols have the same meaning as in Eq. (2.21).

For the capillaries, we have, analogous to Eq. (2.22), the relation

$$f_{5m} = \frac{\varrho L_5}{\pi r_5^2} \frac{\dot{Q}}{n_5} \tag{2.25}$$

It is of interest to compare the relative magnitudes of the viscous and inertial resistances in each of the sets of blood vessels. Neglecting any changes in the apparent blood viscosity, we have, from Eqs. (2.21) and (2.24), the following useful relations:

$$\frac{f_{iv}}{f_{4v}} = \frac{L_i n_4}{L_4 n_i} \left(\frac{r_4}{r_i}\right)^4 \tag{2.26}$$

and

$$\frac{f_{im}}{f_{1m}} = \frac{L_i}{L_1} \frac{n_1}{n_i} \left(\frac{r_1}{r_i}\right)^2 \tag{2.27}$$

where the viscous resistance f_{4v} in the arterioles has been used as reference for comparison of viscous resistances and the inertial resistance f_{1m} in the aorta has been used for comparison of inertial resistances. These relations apply to the arteries and veins but may also be used for the capillaries provided their length L_5 is doubled to account for the different numerical factor in Eqs. (2.22) and (2.25).

The geometric data listed in Table 2.2 have been used to evaluate the above expressions. The results are given in Table 2.5.

It can be seen that the only sensible contributions to the total viscous resistance of the systemic vascular system are in the microscale vessels consisting of the arterioles, capillaries and venules. It can also be seen that the only sensible contributions to the total inertial resistance arise in the macroscale vessels, with the aorta and large arteries providing the most.

Characteristic Microscale and Macroscale Vessels

The total viscous resistance through the entire systemic system, say f_v^*, is expressible from the above arguments simply as the sum of the viscous resistances through the arterioles, capillaries, and venules. To establish a simple relation for this resistance, we define a characteristic length L_c, a characteristic radius r_c, and a characteristic number n_c of the microscale vessels by the equations

Table 2.5.   Relative Contributions of Viscous and Inertial Resistance in Blood Circulation of Humans. *

Vessels	Viscous	Inertial	Remarks
Aorta	10^{-2}	1.0	macroscale
Large arteries	10^{-2}	0.2	vessels
Small arteries	10^{-2}	0.01	
Arterioles	1.0	10^{-4}	microscale
Capillaries	0.5	10^{-5}	vessels
Venules	0.8	10^{-5}	
Small veins	10^{-3}	10^{-3}	macroscale
Large veins	10^{-3}	10^{-1}	vessels
Venae cavae	10^{-3}	10^{-1}	

*Note: The viscous contributions are relative to that in the arterioles and the inertial contributions are relative to that in the aorta.

$$L_c = L_4 \left[1 + \frac{2L_5}{L_4} \frac{n_4}{n_5} \left(\frac{r_4}{r_5}\right)^4 + \frac{L_6}{L_4} \frac{n_4}{n_6} \left(\frac{r_4}{r_6}\right)^4 \right]$$

$$r_c = r_4$$
$$n_c = n_4$$

With these definitions, we then find the total viscous resistance expressible in terms of this characteristic microscale vessel as

$$f_v^* = \frac{4\mu L_c}{\pi r_c^4} \frac{Q}{n_c} \tag{2.28}$$

Alternatively, since the total flow is equal to the product of the blood velocity V_1 in the aorta and the aortic cross section πr_1^2, we may write this relation in the form

$$f_v^* = \frac{4\mu}{n_c} \frac{L_c}{r_c^2} \left(\frac{r_1}{r_c}\right)^2 V_1 \tag{2.29}$$

Similarly, the total inertial resistance, say f_m^*, is expressible as the sum of the inertial resistances in the macroscale vessels. Using the lengths, radii, and number of the macroscale vessels, we may therefore define a characteristic length L_a and radius r_a by the equations

$$L_a = L_1 \left[1 + \frac{L_2}{L_1} \frac{n_1}{n_2} \left(\frac{r_1}{r_2}\right)^2 + \dots \right]$$

$$r_a = r_1$$

The total inertial resistance of the vascular bed is then expressible in terms of this characteristic macroscale vessel as

$$f_m^* = \frac{\varrho L_a \dot{Q}}{2\pi r_a^2} \tag{2.30}$$

Or, with $\dot{Q} = \pi r_1^2 \dot{V}_1$, we have, analogous to Eq. (2.29), the relation

$$f_m^* = \frac{1}{2} \varrho L_a \dot{V}_1 \tag{2.31}$$

where $\dot{V}_1$ denotes the acceleration of the blood in the aorta.

Equations (2.29) and (2.31) have been developed using the properties of systemic vascular bed as a guide. We may, however, also expect similar equations to arise from consideration of the pulmonary vascular bed. The magnitudes of the characteristic parameters in the equations will, of course, be different for the two beds.

In connection with design scaling relations to be considered in the next chapter, we assume the ratios of lengths, radii, and numbers of all microscale vessels to be constant for all mammals. We also assume the ratios of all lengths, radii, and numbers among the macroscale vessels to be constant for all mammals. Thus, we assume the ratios L_5/L_4, n_4/n_5, r_4/r_5, etc., associated with the microscale vessels and the ratios L_2/L_1, n_1/n_2, r_1/r_2, etc., associated with the macroscale vessels to be independent of mammal size. In this case, all lengths, radii, and number of the microscale vessels must vary, respectively, with the characteristic length L_c, radius r_c, and number n_c as mammal size is varied. Similarly, all lengths and radii of the macroscale vessels must vary, respectively, with the characteristic length L_a and radius r_a. The number of macroscopic vessels must also remain fixed as mammal size is varied.

It is worth noting that the scaling relations for the microscale vessels need not be the same as those for the macroscale vessels. It is also not necessary that lengths, radii and numbers of the microscale vessels follow the same law, nor that the lengths and radii of the macroscale vessels vary in the same way.

REFERENCES

ATTINGER, E. D. 1973. Structure and function of the peripheral circulation. In *Engineering Principles in Physiology* Vol. II, J.H.U. Brown and D.S. Gain, ed., pp. 3–47. New York: Academic Press.

COKELET, G. R. 1972. The rheology of human blood. In *Biomechanics—Its Foundations and Objectives,* Y. C. Fung, N. Perrone, M. Anliker, eds., pp. 63–103. Englewood Cliffs: Prentice-Hall, Inc.

DAUGHERTY, R. L., and J. B. FRANZINI. 1965. *Fluid Mechanics with Engineering Applications*. New York: McGraw-Hill.

DOWNING, S. E., and E. H. SONNENBLICK. 1964. Cardiac muscle mechanics and ventricular performance: force and time parameters. *Am. J. Physiol.* 207(3): 705–15.

FRIEDMAN, J. J. 1971. Microcirculation. In *Physiology,* E. E. Sellcurt, ed., pp. 259–73. Boston: Little, Brown.

GUYTON, A. C. 1971. *Textbook of Medical Physiology*. Philadelphia: W. B. Saunders.

HILL, A. V. 1938. The heat of shortening and the dynamic constants of muscle. *Proc. Roy. Soc.,* London 126: 136–95.

LLOYD, T. C., JR. 1971. Respiratory gas exchange and transport. In *Physiology,* E. E. Selkurt, ed., pp. 451–70. Boston: Little, Brown.

NOORDERGRAAF, A. 1978. *Circulatory System Dynamics*. New York: Academic Press.

PAPPENHEIMER, J. R., E. M. RENKIN, and L. M. BORRERO. 1951. Filtration, diffusion and molecular sieving through peripheral capillary membranes. *Am. J. Physiol.* 167(1): 13–46.

SAGAWA, K. 1973. The heart as a pump. In *Engineering Principles in Physiology* Vol. II, J.H.U. Brown and D. S. Gann, eds., pp. 101–26. New York: Academic Press.

SCHMIDT-NIELSEN, K. 1964. *Animal Physiology*. Englewood Cliffs: Prentice Hall.

SELKURT, E. E., and R. W. BULLARD 1971. The heart as a pump: mechanical correlates of cardiac activity. In *Physiology*. E. E. Selkurt, ed., pp. 275–95. Boston: Little, Brown.

SONNENBLICK, E. H. 1962. Force-velocity relations in mammalian heart muscle. *Am. J. Physiol.* 202(5): 931–39.

———. 1964. Series elastic and contractile elements in heart muscle: Change in muscle length. *Am. J. Physiol.* 207(6): 1330–38.

3

Design Theory and Similarity Requirements

Consistent with engineering practice, we now use the results of the previous chapter to devise a theory for the cardiovascular system that will bring together the various parameters characterizing its design and performance. The goal in constructing such a theory is to capture the essentials of the system without unnecessary detail so that realistic relationships between the system variables can readily be identified. In this way, we can gain valuable insight into the workings and design restraints of the system.

Previous works on mathematical modeling of the cardiovascular system, and, in particular, the pumping action of the heart, have frequently envisioned the ventricles as being elastic containers with time-varying elastic parameters connecting ventricular pressures and volumes (see, for example, Sagawa, 1973; Noordergraaf, 1978). The parameters are imagined to change as a result of electrical stimulation of the cardiac muscle, thus accounting for contraction of the ventricles and ejection of blood into the connecting aortic and pulmonary vessels. Outflow is further considered to be impeded by viscous and inertial resistances, and, in this way, reasonable operational models of the system are established.

The design theory to be developed here parallels this general approach. However, it differs substantially in detail in that use is made of the engineering description of cardiac muscle mechanics developed in Chapter 2. As will be seen, this description allows a concise representation of cardiac performance and vascular resistance that will prove very useful in studying the design characteristics and scaling laws of the system.

3.1. ENGINEERING DESIGN MODEL

We saw in Chapter 2 that the flow resistance in the larger blood vessels is mainly inertial in character and can be represented by flow in single characteristic macroscopic vessels, or characteristic connecting vessels, running to and from the left and right sides of the heart. We have also seen that the flow resistance in the smaller blood vessels of the circulation is mainly viscous in character and can be represented by flow in characteristic systemic and pulmonary microscale beds, or characteristic capillary beds, as we shall call them here.

With these ideas before us, we may consider for design purposes the simple model of the cardiovascular system illustrated in Fig. 3.1. The idealized system is seen to consist of left and right heart pumps, with single characteristic connecting vessels from the ventricles to corresponding

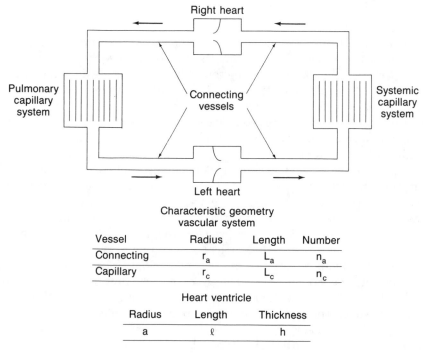

Fig. 3.1 Simplified representation of cardiovascular system for basic design considerations

characteristic capillary beds and from these onward to the adjacent atria reservoirs.

The ventricles of the left and right heart are assumed representable as circular cylinders, with pumping action arising from radial contractions and expansions. Input and output valves are assumed to control the direction of blood flow and allow positive pressure to be maintained in the vascular system, as in the actual case.

During contraction of the ventricles, the output valves of each are open and the input valves closed. Blood is pumped from the left ventricle through the characteristic single vessel, through the characteristic systemic capillary bed, and into the atrium of the right heart. At the same time, blood is also pumped from the right ventricle through its characteristic single vessel, through the characteristic pulmonary bed and into the atrium of the left heart.

During expansion of the ventricles, output valves are closed and input valves are open, and blood is assumed to flow from the atria to the adjacent ventricles of each side of the heart. Near the end of the expansion, we assume, as in the actual response, that atrial contractions occur to complete the expansion and filling of the ventricles. Because of the atrial contractions, we therefore do no need to consider the details of the motion associated with the expansion phase and may assume at the beginning of each ventricular contraction that the ventricle is fully expanded and filled. Also, because the flow from one side of the heart terminates in the atrium of the other, the flows from the left and right sides of the heart are independent provided filling is achieved, and we may therefore consider either side in formulating a mathematical description of the system.

Ventricular Equation

We denote the internal pressure in a ventricle by P, the hoop force per unit length in the walls by F' (see Fig. 3.2) and the inward radial displacement by U. Neglecting inertia of the ventricle in comparison with pressure forces, as can be shown to be acceptable, we may relate these variables using the force-balance equation:

$$P(a - U) = F' \tag{3.1}$$

where a denotes the inside radius of the ventricle about which the radial displacement occurs.

Hoop-Force Equation

For pumping action, we assume cyclic contractile hoop forces, per unit length, about a fixed value. The total hoop force then consists of this

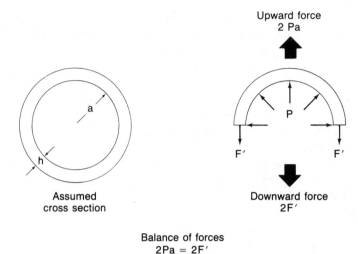

Fig. 3.2 Representation of balance of forces per unit of length acting on a ventricle

force and an elastic force associated with the radial displacement as described in Chapter 2. We accordingly represent the cyclic force by the general expression $F_oG(\omega t)$ where F_o denotes the amplitude of the force and $G(\omega t)$ denotes an appropriate cyclic function of ωt having unit amplitude, with ω denoting frequency and t denoting time. The elastic force is likewise represented by a general expression of the form $EhH(U/a)$, where E denotes an elastic modulus of the ventricular wall, h denotes the wall thickness, and $H(U/a)$ denotes an appropriate function describing the variation of the elastic force with relative displacement, or strain, U/a. With F_s representing the isometric, or fixed, part of the contractile force, we then have the hoop force expressible from Eq. (2.18) as

$$F' = F_s + F_o G(\omega t) - EhH(U/a) \tag{3.2}$$

Pressure Equation

The internal pressure in the ventricle is equal to the sum of the static pressure and the pressure losses from inertial and viscous resistance. If P_s denotes the static pressure, we thus have the equation

$$P = P_s + f_m^{\cdot} + f_v^{\cdot} \tag{3.3}$$

where $f_m^{\cdot}$ and $f_v^{\cdot}$ denote, respectively, the inertial and frictional losses.

The characteristic length and radius of the connecting vessels are denoted by L_a and r_a, respectively. In the characteristic capillary system, we assume a large number n_c of small vessels of length L_c and radius r_c. From our earlier discussion in Chapter 2, we accordingly have the pressure losses expressible as

$$f_m^{\cdot} = \frac{1}{2} \varrho L_a \dot{V}_1 \tag{3.4}$$

and

$$f_v^{\cdot} = 4\mu \frac{1}{n_c} \frac{L_c}{r_c^2} \left(\frac{r_a}{r_c}\right)^2 V_1 \tag{3.5}$$

where ϱ and μ denote, respectively, the density and viscosity of the blood, and V_1 and $\dot{V}_1$ denote, respectively, the velocity and acceleration of the blood in the connecting tubes. Combining Eqs. (3.3) through (3.5), and defining the symbol η as

$$\eta = \frac{1}{n_c} \frac{L_c}{r_c^2} \left(\frac{r_a}{r_c}\right)^2 \tag{3.6}$$

we find the pressure expressible as

$$P = P_s + \frac{1}{2} \varrho L_a \dot{V}_1 + 4\mu\eta V_1 \tag{3.7}$$

The velocity and acceleration of the blood in the connecting vessel are related to the radial inward velocity U and inward acceleration $\dot{U}$ of the ventricular wall through the condition of mass conservation and are determined by the expressions

$$V_1 = \frac{2a\ell}{r_a^2} \left(1 - \frac{U}{a}\right) \dot{U} \tag{3.8}$$

and

$$\dot{V}_1 = \frac{2a\ell}{r_a^2} \left(1 - \frac{U}{a}\right) \ddot{U} - \frac{2\ell}{r_a^2} \dot{U}^2 \tag{3.9}$$

where ℓ denotes the length of the ventricle.

Substituting Eqs. (3.8) and (3.9) into relation (3.7) and multiplying through by the radius a of the ventricle, we then find the pressure force Pa expressible in terms of the ventricular motion as

$$Pa = P_s a + M\lambda\ddot{U} - M'\dot{U}^2 + C\lambda\dot{U} \tag{3.10}$$

where $\lambda = 1 - U/a$ and M, M', and C are defined by the equations

$$M = \varrho L_a \frac{a^2\ell}{r_a^2}, \ M' = \frac{M}{a}, \ C = 8\mu\eta \frac{a^2\ell}{r_a^2} \tag{3.11}$$

Governing Equation

Assuming that ventricular outflow begins at the end of isometric contraction, the static pressure term $P_s a$ in Eq. (3.10) must equal the static force term F_s in Eq. (3.2). On combining Eqs. (3.1), (3.2), and (3.10), we then find the equation for the radial displacement of the ventricle to be expressible as

$$M\lambda^2\ddot{U} - M'\lambda\dot{U}^2 + C\lambda^2\dot{U} - P_s U = F_o G(\omega t) - EhH(U/a) \tag{3.12}$$

In connection with this equation, we may observe that, since the ventricular contractions are periodic in nature, the displacement will equal an amplitude value, say U_o, at the maximum contraction when the velocity of contraction equals zero. Thus, according to the design theory, this last equation is that governing the periodic ventricular contractions of either side of the heart, subject to the initial (and periodically repeating) conditions that $U = U_o$ and $\dot{U} = 0$.

Of course, while Eq. (3.12) is of the same form for either side of the heart, the values of the variables will generally be different because of the different vascular beds of the two sides and the different requirements on the ventricular dimensions.

3.2. SIMILARITY REQUIREMENTS

Having Eq. (3.12) before us, we may now establish conditions that must apply if, as indicated in Chapter 1, similar response of cardiovascular variables is to exist among mammals of different size. To do this, we may again use methods of dimensional analysis such as described in Chapter 1. However, unlike this earlier treatment, which provided no appropriate similarity conditions, we now have available the above equation to indicate the basic groupings of variables governing the system. This accordingly allows a more specialized treatment than was possible in Chapter 1 where our limited information allowed consideration only of individual variables.

From Eq. (3.12) and the associated initial (periodic)·conditions, we now see, in particular, that the ventricular displacement U must depend on the grouped variables M, M', and C, defined by Eq. (3.11), as well as on the variables P_s, F_o, Eh, ω, t, U_o, and a. Noticing further from the definition of M' as M/a that this variable may be excluded from the list as long as M and a are included, we may express this dependence mathematically in terms of appropriate dimensionless ratios as

$$\frac{U}{U_o} = f\left(\frac{MU_o\omega^2}{F_o}, \frac{CU_o\omega}{F_o}, \frac{Eh}{F_o}, \frac{P_sU_o}{F_o}, \frac{U_o}{a}, \omega t\right) \tag{3.13}$$

where $f(-)$ denotes a general function of the indicated dimensionless variables.

Now let us consider this relation in some detail. If conditions are such that the first five ratios on the right-hand side of this equation are fixed for all mammals, regardless of size, we see that the equation indicates simply that the ratio U/U_o will vary only with the time variable ωt and, hence, will itself be independent of mammal size. In this case, similar response of ventricular displacement during contraction will then exist for all mammals. At any fixed time ωt in the contraction cycle of the ventricle, the radial displacement U will be proportional to its amplitude U_o, regardless of mammal size.

If similar response exists for the ventricular displacement, it will, of course, also exist for other related variables. We therefore see from the above argument that the conditions for similar behavior of the cardiovascular system of mammals is that the following ratios be constant for all mammals:

$$\pi_1 = \frac{MU_o\omega^2}{F_o}, \quad \pi_2 = \frac{CU_o\omega}{F_o}$$

$$\pi_3 = \frac{Eh}{F_o}, \quad \pi_4 = \frac{P_sU_O}{F_o}, \quad \pi_5 = \frac{U_o}{a} \tag{3.14}$$

In addition to these relations, we assume similar construction materials for all mammals such that mass densities, elastic moduli, and blood viscosity are all essentially independent of size. We also use the fact that the total weight of the blood and the weight of the heart itself are each proportional to animal weight, consistent with known similarity of animals, as discussed in Chapter 1. Since for size independent densities, weights are proportional to volumes, these conditions give the following relations to be satisfied for each side of the heart:

$$B_r + \pi r_a^2 L_a + \pi n_c r_c^2 L_c + \pi a^2 \ell \; \alpha \; W \tag{3.15}$$

and

$$2\pi a^2 h + 2\pi a \ell h \ \alpha \ W \tag{3.16}$$

where B_r denotes the volume of blood in the atrium, W denotes animal weight and, as earlier, the symbol α denotes proportionality.

Now, each of the products in relations (3.15) and (3.16), like their sums, may be assumed proportional to animal weight. From these relations, we then see that all ventricular dimensions must scale in the same way, namely, according to the relations

$$a \ \alpha \ W^{1/3}, \ \ell \ \alpha \ W^{1/3}, \ h \ \alpha \ W^{1/3} \tag{3.17}$$

Using these relations and the last three of Eqs. (3.14), with π_3, π_4, and π_5 constant, we also find, with the help of the assumption of scale-independent elastic moduli, the scaling relations

$$U_o \ \alpha \ W^{1/3}, \ F_o \ \alpha \ W^{1/3}, \ P_s \ \alpha \ W^0 \tag{3.18}$$

where W^0 means there is no dependence on weight or size.

Since these last relations show that both U_o and F_o vary with animal weight raised to the same power, they are proportional under change of scale. The first two of Eqs. (3.14), with π_1 and π_2 constant, then give, with the help of relations (3.17) and (3.18) and the definitions of M and C and the assumption of scale-independent density and viscosity, the additional requirements

$$L_a \ \alpha \ \omega^{-1}, \ \eta \ \alpha \ W^o \tag{3.19}$$

where η denotes the geometric factor associated with the capillary system, as given by Eq. (3.6). We also have from relation (3.15) the following conditions on the vascular system:

$$r_a^2 L_a \ \alpha \ W, \ n_c r_c^2 L_c \ \alpha \ W \tag{3.20}$$

Finally, from relation (3.20) and the above requirement on the geometric factor η, we see that the length L_a of the connecting vessels and the length L_c and number n_c of the very small vessels in the capillary system must scale according to the relations

$$L_a \ \alpha \ W r_a^{-2}, \ L_c \ \alpha \ W^{1/2} r_c r_a^{-1}, \ n_c \ \alpha \ W^{1/2} r_a r_c^{-3} \tag{3.21}$$

In summary, then, we have found from the present design theory the following important scaling laws: (a) the relations (3.17) showing that all linear dimensions of the ventricles (and, more generally, all linear dimensions of the heart) must scale with animal weight raised to the one-third power; (b) the relations (3.18) showing that the amplitudes of the ventricular contractions and hoop force per unit length must scale with animal weight raised to the one-third power, and that the static ventricular pressure must be independent of animal size; (c) the first of relations (3.19) showing that the length of connecting vessels and the heart rate must be inversely proportional to one another; and (d) the three scaling relations (3.21) restricting the variation of the geometry of the vascular system with animal size.

3.3 ADDITIONAL DESIGN ASSUMPTIONS

Since the above three geometric scaling relations (3.21) connect five unknown quantities $(L_a, r_a, n_c, L_c, r_c)$, we must have two additional relations between them if we are to solve for the scaling relations of each. To obtain these relations, we examine here additional fundamental processes governing the design of the cardiovascular system.

We first consider the basic cellular structure of the heart. We assume the number of tissue cells n_s in the heart to be proportional, under size change, to the number of capillaries present. For constant tissue density, or weight per unit volume, the characteristic volume of an average heart cell will then be proportional to the ratio of heart weight to number of capillaries. We also assume the number of capillaries in the heart to be proportional, under size change, to the number n_c in the entire system. Since heart weight is proportional to animal weight, we may then express the characteristic length ℓ_s of an average heart cell, or equivalently, an average body cell, as

$$\ell_s \propto (W/n_c)^{1/3} \qquad (3.22)$$

where, in writing this expression, all cell dimensions have been considered to vary with animal size in the same way.

We now consider the rate of cardiac contractions and observe that the time between the initiation of the action potential in the upper right atrium (the S-A node) and the onset of ventricular contraction must be directly proportional to the distance between the points involved and inversely proportional to the speed at which the signal propagates. Moreover, for similar heart response among mammals of different size, the period between heartbeats must be proportional to this time.

Now, the distance over which the signal propagates may be considered to be proportional under change of size to any characteristic heart dimension such as the ventricular length ℓ introduced earlier. In addition, it has long been recognized that conduction speed in the heart is related to the diameter of the fibers carrying the signal, with larger fibers having greater conduction speeds than smaller ones (see, for example, Guyton 1971). If we assume this relation to be of simple power-law form, we may thus express the conduction speed C_o and the period T between heartbeats as

$$C_o \propto d^b, \ T \propto \ell/d^b \tag{3.23}$$

where d denotes fiber diameter and b denotes a positive exponent.

An appropriate value for the exponent b in the above relations is not well established. Idealized electrical theory of cables suggests a value of 0.5, while measurements on nerve fibers show values between 0.5 and 1.0 (Jack, Noble, and Tsien 1975). In the present work, we assume, tentatively, a design value of 0.67, that is, two-thirds. We shall see later that this assumption does, in fact, lead to the known scaling relations for heart rate and oxygen-consumption rate identified in Chapter 1. We also note that cardiac fiber consists of a series connection of single cells, so that fiber diameter is the same as the characteristic cell length described by relation (3.22). Thus, relations (3.23) become expressible as

$$C_o \propto d^{2/3}, \ T \propto \ell \, (W/n_c)^{-2/9} \tag{3.24}$$

Finally, substituting into this last relation the scaling law for ℓ from relations (3.17) and using the fact that the heart rate ω is the inverse of the heart period T, we have the design relation

$$\omega \propto W^{-1/9} \, n_c^{-2/9} \tag{3.25}$$

We next consider the mass m of a substance diffusing into or out of a cardiac cell, or average body cell, during a cycle proportional to the cardiac pumping cycle. Mathematically, we may express this in general functional form as

$$m = m \, (\ell_s, D, \omega, \Delta C) \tag{3.26}$$

where D denotes the usual diffusion coefficient (in units of area per unit time) defined by Fick's law of diffusion (Appendix D) and ΔC denotes the characteristic difference between concentrations of the substance inside and outside the cell (in units of mass per unit volume). Using dimensional reasoning, we may write this result as

$$\frac{m}{\Delta C \ell_s^3} = f \ \frac{\omega \ell_s^2}{D} \tag{3.27}$$

where $f(-)$ denotes a function of the indicated variable.

With ΔC and D constant, we thus see that for the diffused mass to be proportional to the cell volume, as required for similar cell processes among mammals, the product $\omega \ell_s^2$ must be constant, independent of animal size. This result can also be obtained directly from Fick's law of diffusion and was arrived at earlier in connection with cardiac excitation [Eq. (2.6)] by assuming a similar process. Using the previous relation for ℓ_s, we accordingly have the following proposed second fundamental relation governing the design of the cardiovascular system:

$$\omega \; \alpha \; W^{-2/3} \, n_c^{2/3} \qquad (3.28)$$

3.4 SCALING LAWS

We may now use the above two design assumptions with the scaling relations (3.21) to establish the geometric scaling laws governing the cardiovascular design. For this purpose, we may use the first of relations (3.19) to eliminate the heart rate in relations (3.25) and (3.28). These can then be solved to give the scaling relations for n_c and L_A. With these, the scaling relations (3.21) can then be used to solve for the relations governing the remaining geometric variables.

In the above manner, we therefore derive the scaling relations for the radius and length of the characteristic connecting vessels of the system. Including the scale-independent number n_a of these vessels, we may write these as

$$r_a \; \alpha \; W^{3/8}, \; L_a \; \alpha \; W^{1/4}, \; n_a \; \alpha \; W^0 \qquad (3.29)$$

The scaling relations for the radius, length, and number of the characteristic capillary vessels are likewise determined as

$$r_c \; \alpha \; W^{1/12}, \; L_c \; \alpha \; W^{5/24}, \; n_c \; \alpha \; W^{5/8} \qquad (3.30)$$

From relation (3.22) and the discussion leading to it, we also then find the scaling relations for the characteristic length of a cardiac cell, or average body cell, and the number of such cells as

$$\ell_s \; \alpha \; W^{1/8}, \; n_s \; \alpha \; W^{5/8} \qquad (3.31)$$

The above relations describe geometric scaling laws for the cardiovascular system and the tissue cells. With these, the physical scaling law governing the heart rate can now be determined from relations (3.25) or (3.28), or from the first of relations (3.19), as

$$\omega \; \alpha \; W^{-1/4} \qquad (3.32)$$

Moreover, since the mass of a substance consumed (or expelled) by an average body cell during a cycle of operation is proportional to the characteristic cell volume ℓ_s^3, as determined from relations (3.22), the total mass for the entire body is proportional to the corresponding total cell volume $n_s \ell_s^3$. The net rate of consumption is therefore proportional to $\omega n_s \ell_s^3$ which, with relations (3.22) and (3.28), is seen to be proportional simply to $n_s \ell_s$. Using the scaling relations for n_s and ℓ_s, we then see that the oxygen-consumption rate Q_o must obey the relation

$$Q_o \alpha W^{3/4} \tag{3.33}$$

These last two scaling relations are, of course, in precise agreement with the empirical scaling laws for heart rate and oxygen-consumption rate that were described in Chapter 1. Thus, we see that the present theory not only provides us with geometric scaling laws for the cardiovascular system and average tissue cells, but also provides us with two very important physical scaling laws describing the system performance. We see, in addition, that the origin of the physical scaling relation for oxygen-consumption rate rests in fundamental diffusion processes occurring across cell surfaces and in the specialized manner in which the number and linear dimensions of average body cells must vary with animal weight. The scaling relation for the heart rate likewise is such that the amount of a substance diffusing into or out of a cardiac cell, or, equivalently, an average body cell, during a cycle of operation is directly proportional under change of scale to the cell volume. The actual electrical discharge responsible for the heart rate is also consistent with this process as shown earlier by Eq. (2.6).

It is interesting to notice that, in the present development, we could replace the design conditions (3.25) and (3.28) with the empirical scaling relations for heart rate and oxygen-consumption rate described in Chapter 1 and still obtain the geometric scaling relations found above. Of particular significance here, however, is the fact that these scaling laws have now been derived from fundamental design considerations.

Although not explicit in the above development, it can be seen on inspection that relations (3.25) and (3.28) determine directly the scaling relations for heart rate and number of capillaries n_c. Relation (3.22) then determines the scaling law for the characteristic cell size ℓ_s, and the product $n_s \ell_s$ (with n_s proportional to n_c) determines the scaling law for oxygen-consumption rate. These scaling results are thus independent of the additional relations (3.21) which define the scaling of the geometry of the vascular system of mammals. To the extent that relations (3.22), (3.25), and (3.28) apply to animals other than mammals, we may therefore expect similar scaling laws for their heart rate and oxygen-consumption rate.

We may recall that, in the development of relation (3.25), the assumption was made that the exponent b in the expression relating conduction speed to fiber diameter was chosen equal to two-thirds. If, instead of this choice, we assume a value of one-half, consistent with that suggested by idealized electrical theory of cables, we then find, for example, that the scaling relations for heart rate and oxygen-consumption rate involve animal weight raised to the -0.266 and 0.733 powers, respectively. These relations are not appreciably different from the reciprocal one-fourth and three-fourths relations developed above using the two-thirds value for the exponent. The scaling relations are therefore not particularly sensitive to the exact value of the exponent chosen. The exponent value of two-thirds is preferred here simply because it gives precisely the reciprocal one-fourth and three-fourths relations identified in Chapter 1 from experimental measurements.

We may also note that the present theory is consistent with the proposed similarity principle of Noordergraaf, Li, and Campbell (1979) for pressure waves in the aorta at frequencies of the order of the heart rate or greater. According to this principle, the ratio of the length of a wave in this frequency range to the aortic length is constant, independent of animal size. The result is based on the known condition that the wave speed is constant for the frequencies considered. The wave length is then inversely proportional to the heart rate. The ratio of wave length to aortic length is thus inversely proportional to the product of the heart rate and this latter length. But this product is constant according to the first of relations (3.19), and, hence, the ratio of wave length to aortic length is also constant, in agreement with the proposed similarity principle.

The scaling relations (3.29) through (3.33) can be illustrated by assuming typical values for the human and calculating corresponding values for other mammals. If, for example, we assume a typical capillary radius of 0.005 mm for a human weighing 70 kg, we may calculate the corresponding value for a mammal of weight W using the first of relations (3.30) in the form

$$r_c = k_r W^{1/12}$$

where k_r denotes a constant which may be determined from the assumed values for the human as

$$k_r = 0.00351 \text{ mm/kg}^{1/12}$$

Alternatively, we may write the relation as

$$\frac{r_c}{0.005} = \left(\frac{W}{70}\right)^{1/12}$$

and calculate values of r_c for specified values of W without explicitly evaluating the constant k_r.

Various such calculations have been made and tabulated in Table 3.1 for the geometry of large arteries, capillaries, average tissue cells, heart rate, and oxygen-consumption rate of the mouse and the elephant. As expected from the form of the geometric scaling relations, the capillary radius is found to vary least, and the characteristic number of the capillaries most, in scaling from the mouse to the elephant. As seen in Chapter 1, the heart rate of the mouse is more than 7 times that of the human and more than 18 times that of the elephant. Also, the oxygen-consumption rate of the mouse is almost 400 times smaller than that of human and 6500 times

TABLE 3.1. Calculated Values of Cardiovascular Parameters for Mouse and Elephant from Assumed Values for Humans

Quantity	Mouse (0.025 kg)	Human* (70 kg)	Elephant (3000 kg)
LARGE ARTERIES			
Radius r_a	0.15 mm	3 mm	12 mm
Length L_a	34 mm	250 mm	640 mm
Number n_a	50	50	50
CAPILLARIES			
Radius r_c	0.0026 mm	0.005 mm	0.0068 mm
Length L_c	0.19 mm	1.0 mm	2.2. mm
Number n_c	2.1×10^7	3×10^9	3.1×10^{10}
AVERAGE BODY CELL			
Linear dimension ℓ_s	0.004 mm	0.010 mm	0.016 mm
HEART RATE			
Rate ω	509 beats/min.	70 beats/min.	27 beats/min.
OXYGEN-CONSUMPTION RATE			
Rate Q_o	0.62 ml/min.	240 ml/min.	4020 ml/min.

* Assumed values.

smaller than that of the elephant. In spite of these wide variations, the basic design equations that apply to the mouse likewise apply to the elephant.

REFERENCES

GUYTON, A. C. 1971. *Textbook of Medical Physiology*. Philadelphia: W. B. Saunders Company.

JACK, J. J. B., D. NOBLE, and R. W. TSIEN. 1975. *Electric Current Flow in Excitable Cells*. Oxford: Clarendon Press.

NOORDERGRAAF, A. 1978. *Circulatory System Dynamics*. New York: Academic Press.

NOORDERGRAAF, A., J. K. J. LI, and K. B. CAMPBELL. 1979. Mammalian hemodynamics: a new similarity principle. *J. Theor. Biol.* 79: 485–89.

SAGAWA, K. 1973. Comparative models of overall circulatory mechanics. In *Advances in Biomedical Engineering* Vol. 3, J. H. Brown and J. F. Dickson, III, eds., pp. 1–92. New York: Academic Press.

4

Applications of Design Theory

With the design theory of Chapter 3 before us, we may now apply it to the prediction and understanding of the scaling relations governing various physiological variables of mammals of different size. We have already seen that the theory is successful in describing the known scaling relations for heart rate and oxygen-consumption rate of mammals. Our aim here is to apply and extend the scaling theory of Chapter 3 to include additional variables and interpretations.

4.1 GEOMETRIC SCALING RELATIONS

We shall consider first some examples illustrating the validity of the geometric scaling relations established in Chapter 3. We recall that the ventricular geometry assumed in the theoretical model of Chapter 3 was cylindrical, with scaling laws for its length, radius, and wall thickness given by relations (3.17). The fact that all dimensions scale the same way, namely, as animal weight to the one-third power, suggests that the result is more general than that associated only with the assumed cylindrical geometry. This can, in fact, be shown to be the case. Thus, the scaling relations for the ventricle geometry may be expected to apply to measurements made on actual ventricles.

Similarly, the scaling relations (3.29) and (3.30) were established for characteristic connecting vessels and capillaries of the theoretical model. However, from scaling arguments given in Chapter 2, they may also be expected to apply, in the first case, to the main arteries and veins of the vascular system, and, in the second case, to the arterioles, capillaries, and venules of the system. Keeping in mind this generality, we will continue to refer, when convenient, to the arteries and veins as connecting vessels, and to the arterioles, capillaries, and venules as characteristic capillary vessels.

Ventricular Geometry

For a first example, we examine the scaling relations (3.17) requiring that linear dimensions of the heart ventricles vary with animal weight to the one-third power. We may examine this prediction using measurements of the length l of the left-ventricle cavity of the heart of various mammals, as made by Clark (1927). The measurements are given in Table 4.1, together with corresponding values for the animal weights W. In order to express the scaling relation in equation form, values of $\ell W^{-1/3}$ have been calculated and the average determined as 1.78. The resulting scaling equation then takes the form

$$\ell = 1.78 \, W^{1/3} \tag{4.1}$$

This equation has been graphed in Fig. 4.1 using logarithmic scale axes as discussed in Chapter 1. The individual measurements have also been plotted in this figure, and it can be seen that the correlation is, indeed, remarkably good. In fact, a best-fit analysis of the ventricle-length and animal-weight data in Table 4.1 to an assumed power-law expression provides an exponent of 0.33, in excellent agreement with the theoretical one-third value. This result is consistent with the initial observation by Clark that ventricular length varied approximately with the cube root of heart weight. However, since his studies did not reveal a precisely propor-

---

TABLE 4.1 Measurements of Cavity Length ℓ of Left Ventricle and Animal Weight W by Clark (1927)

Animal	W (kg)	ℓ (cm)
Mouse	0.025	0.55
Rat	0.16	1.0
Rabbit	2.0	2.2
Dog	12	4.0
Sheep	45	6.5
Ox	460	12
Horse	610	16

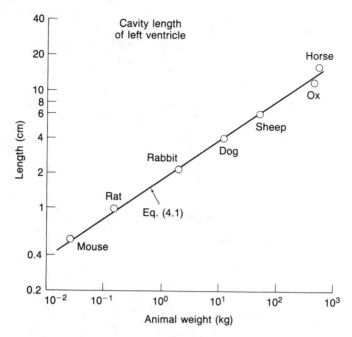

Fig. 4.1   Comparison of measurements in Table 4.1 for the left-ventricle cavity with predictions from the theoretical one-third scaling law in the form of Eq. (4.1)

tional relation between heart weight and animal weight, the connection described by Eq. (4.1) was not given.

More recently, Holt, Rhode, and Kines (1968) provided detailed measurements of the end-diastolic volume B_m of the ventricles of a variety of mammals. Typical values for the left ventricle are listed in Table 4.2. According to the relations (3.17), the ventricular volume should scale directly with animal weight. That this is the case can be seen in Fig. 4.2, where the measurements have been plotted against animal weight along with predictions from the simple proportional relation proposed by Holt et al., namely

$$B_m = 2.25W \tag{4.2}$$

The data are seen to agree well with the predictions, thus providing additional evidence that the ventricular geometry scales in accordance with relations (3.17). Similar results also apply to the right ventricle.

It is interesting to consider the ratio B_m/ℓ using Eqs. (4.1) and (4.2) and thereby determine the scaling relation for the average cross-sectional area A_m of the ventricle. The result is expressible as

$$A_m = 1.24W^{2/3} \tag{4.3}$$

TABLE 4.2. **Measurements of End-Diastolic Volume**
B_m of Left Ventricle and Animal Weight
W by Holt et al. (1968)

Animal	W (kg)	B_m (ml)
Rat	0.49	1.4
Rabbit	3.7	3.4
Dog	18	71.1
Goat	42	97.9
Sheep	77	193
Pig	188	208
Horse	466	987
Cow	518	868

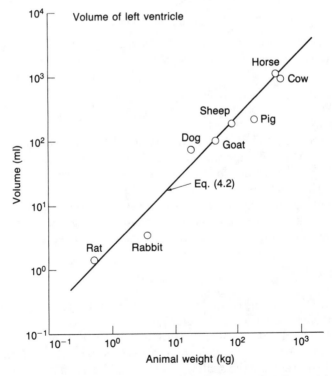

Fig. 4.2 Comparison of measurements of Table 4.2 for the left-ventricle volume with predictions from the theoretical linear scaling law in the form of Eq. (4.2)

where A_m is in units of cm^2. For the human, weighing 70 kg, this equation gives 21 cm^2. If, as in our theoretical model in Chapter 3, we think of the ventricle as being cylindrical in shape, this area corresponds to a radius of about 2.6 cm. The corresponding length of the cylinder is about 7.3 cm.

Aortic Radius

For a next example, we consider the three-eighths scaling law for the radius of the characteristic connecting vessels (arteries and veins) of the cardiovascular system, as given by the first of relations (3.29). This relation can be expected to apply to the aorta of mammals where, in the theory, inertial resistance to blood flow dominates. Fortunately, measurements of the flow area of the aorta of a number of animals have also been made by Clark (1927) and are available for examination. Table 4.3 lists the aortic radius R as calculated from these measurements, together with respective animal weights.

To obtain the coefficient in the theoretical scaling equation, values of $RW^{-3/8}$ have been calculated and the average value found equal to 0.18. Hence, the scaling equation for the aortic radius may be written as

$$R = 0.18\,W^{3/8} \tag{4.4}$$

This equation has been graphed in Fig. 4.3 along with the individual measurements. The agreement is seen to be very good. A best-fit analysis of the radius and body-weight data of Table 4.3 to an assumed power-law expression reveals, in fact, an exponent of 0.41, which is in good agreement with the theoretical three-eighths (0.375) exponent. Again, we may note that Clark (1927) recognized initially that a 0.8 power-law relation existed between his measurements of aortic-flow area and heart weight.

TABLE 4.3.   Values of Aortic Radius R for Animals of Weight W as Determined from Flow Area Measurements of Clark (1927)

Animal	W (kg)	R (cm)
Mouse	0.025	0.04
Rat	0.18	0.08
Rabbit	2.0	0.18
Dog	12	0.60
Sheep	45	0.75
Ox	590	2.0
Horse	610	2.7

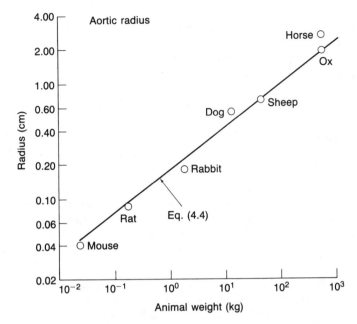

Fig 4.3 Comparison of measurements of aortic radius given in Table 4.3 with predictions from the theoretical three-eighths scaling law in the form of Eq. (4.4)

This corresponds, in turn, to a 0.4 power-law relation for aortic radius with heart weight, in close agreement with the result given here in terms of animal weight.

As a further illustration of the scaling relation for aortic radius, we may apply it to calculate the value expected for the giant sulphur-bottom whale weighing approximately 100,000 kg. In this case, we find from Eq. (4.4) a rounded value of 14 cm which is in remarkably good agreement with the value of 15 cm determined from a corresponding cross-sectional area measurement reported by Clark. Alternatively, and perhaps more dramatically, we may scale directly from the small mouse to the whale using the aortic-radius measurement of 0.04 cm given in Table 4.3. Thus, we have

$$\frac{R}{0.04} = \left(\frac{100,000}{0.025}\right)^{3/8}$$

giving a rounded value of $R = 12$ cm. This may also be considered to be in remarkably good agreement with the measured value of 15 cm when it is remembered that the whale weighs four million times more than the mouse and that the scaling is based on a single measurement.

In connection with this last result, we may consider what a similar calculation would give if the scaling law were a conventional one-third power law, like that for ventricular lengths, rather than the three-eighths law of the theory. In this case, we find a value of $R = 6$ cm which is only half that given by the three-eighths law and in very poor agreement with the measurement. In a similar manner, we may scale the value for the mouse in Table 4.3 to that listed for the ox. The three-eighths law gives a value of 1.7 cm in good agreement with the measured value of 2.0 cm, while the one-third relation gives a value of only 1.1 cm. Thus, on the basis of these calculations, we see there can be no question that the three-eighths scaling law of the theory is favored by the measurements over a one-third relation such as applies when all dimensions of a body scale in the same way.

With the results of this section, we see that the measurements of the ventricular geometry and aortic radius of mammals are, indeed, described very well by the scaling relations of the present theory. It will be recalled that the theoretical relation for the ventricular dimensions is based on the condition that the weight and blood volume of the heart vary directly with animal weight. In contrast, the theoretical scaling relation for the radius of the connecting vessels is determined by the full set of similarity conditions applying to the cardiovascular system.

4.2 BLOOD MOTIONS AND PRESSURES

We next consider scaling laws for blood flows, velocities, and pressures in the characteristic connecting vessels (arteries and veins) and in the characteristic capillary vessels (arterioles, capillaries, and venules) of the system. In this regard, it will be convenient to keep in mind the similarity relation of Chapter 3 requiring relative ventricular displacements to be independent of animal size for proportional times in the heart cycle. With this condition, for example, the ratio of instantaneous blood flow to average blood flow will be the same for all mammals at proportional times in the heart cycle. As an application of the scaling law for blood velocities, we also develop in the present section the scaling law for total circulation time of an elemental volume of blood.

Cardiac Output

We first establish the scaling relation for cardiac output, that is, the average blood flow out of the heart. Because of the similarity that exists in ventricular response over a cycle of pumping, the cardiac output will be proportional to the product of the heart rate ω and initial volume of the ventricle B_m. Since the latter is proportional to animal weight and the former proportional to animal weight to the negative one-fourth according to relations (3.17) and (3.32), we thus have the cardiac output Q_b expressible as

$$Q_b \ \alpha \ \omega B_m \ \alpha \ W^{3/4} \qquad (4.5)$$

that is, as animal weight to the three-fourths power. This is the same type of power law as that predicted by the theory for oxygen-consumption rate, and we see that the theory requires that the ratio of oxygen-consumption rate to cardiac output be constant, independent of scale. Such a condition was noted in Chapter 1 where the scaling law for the cardiac output was established from that of the oxygen-consumption rate using a proportional factor of 100 ml of blood delivered for every 5 ml of oxygen consumed.

The validity of this scaling relation has been confirmed by Holt, Rhode, and Kines (1968) who measured cardiac output of various anesthetized mammals and showed that the measurements were in close agreement with a three-fourths scaling relation. Assuming this relation, the scaling equation developed from the data is expressible as

$$Q_b \ = \ 209 \ W^{3/4} \qquad (4.6)$$

where Q_b is in units of ml/min. and W is in kilogram units. Interestingly, the coefficient in this equation is only about 7% less than that in the similar equation for cardiac output given in Chapter 1 and derived from oxygen-consumption rate measurements.

To illustrate how well the measurements follow the predictions of Eq. (4.6), typical average values have been collected in Table 4.4. For the cow of weight 518 kg, the equation predicts a cardiac output of 22,690 ml/min. which is in good agreement with the measured value of 23,900 ml/min. Similarly, for the rat weighing 0.49 kg, the equation gives a value

TABLE 4.4. Typical Measurements of Cardiac Output (Q_b) and Animal Weight (W) by Holt et al. (1968)

Animal	W (kg)	Q_b (ml/min.)
Rat	0.49	120
Rabbit	3.7	380
Dog	18	2100
Goat	42	4000
Sheep	77	5200
Pig	188	6700
Cow	518	23,900

for cardiac output of 122 ml/min. which is in very close agreement with the measured value of 120 ml/min.

These and other comparisons are shown in Fig. 4.4 where predicted and measured values of cardiac output are plotted against animal weight using logarithmic-scale axes. The agreement is seen to be excellent, with all the measurements following closely the straight line representing Eq. (4.6).

Blood Motions and Pressures in Connecting Vessels

Noting again the similarity in pump behavior that exists over the entire heart cycle, we may next observe that the scaling relation (4.5) for cardiac output also applies to the instantaneous blood flow, say Q, at relative times in the cycle. Moreover, the instantaneous blood velocity V_a in any specified class of arteries or veins is equal to the blood flow divided by the net cross-sectional area of the vessels. The latter is proportional to the product of the number n_a of such vessels and the square of the tube radius r_a. From relation (3.29), this can be seen to be proportional to animal weight raised to the three-fourths power. This is the same scaling law as the blood flow, so we have the result that

$$V_a \ \alpha \ Q/n_a r_a^2 \ \alpha \ W^o \qquad (4.7)$$

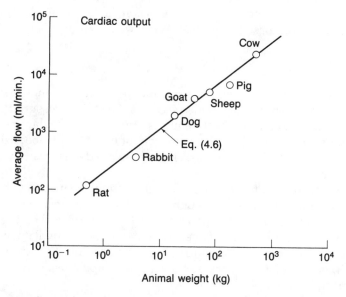

Fig. 4.4 Comparison of measurements of cardiac output given in Table 4.4 with predictions from the theoretical three-fourths scaling law in the form of Eq. (4.6)

that is, that the instantaneous blood velocities in the arteries and veins at relative times in the pumping cycle are independent of animal size.

Because of the similarity in the pumping cycle, the value of the instantaneous acceleration $\dot{V}_a$ of the blood in the arteries and veins will be proportional under size change to the product of the heart rate ω and the associated instantaneous blood velocity. Thus, with the above relation, we have

$$\dot{V}_a \ \alpha \ \omega V_a \ \alpha \ W^{-1/4} \qquad (4.8)$$

where the scaling relation (3.32) for heart rate has also been used here.

From these last relations, we see, for example, that the maximum and average blood velocities in the aorta of, say, the human and the mouse are the same, but that the maximum and average blood accelerations are about seven times greater in the mouse than in the human.

Using results from Chapter 3, we can also establish the theoretical result that the ventricular pressure P is independent of animal size. In particular, the increase in this pressure over the mean value P_s is expressible in terms of the blood acceleration $\dot{V}_1$ and velocity V_1 in the connecting vessel using Eq. (3.7) in the form

$$P - P_s = 1/2 \ \varrho \ L_a \dot{V}_1 + 4\mu\eta V_1$$

where, as earlier, L_a denotes the characteristic length of the connecting vessel and ϱ, μ, and η denote constants defined in Chapter 3. With relations (4.7) and (4.8) above and the condition that ω is inversely proportional to L_a, as required by relation (3.19), we see from this last equation that

$$P - P_s \ \alpha \ W^o \qquad (4.9)$$

that is, that the pressure difference is independent of animal size. From the third of relations (3.18), we also see that the static pressure P_s is independent of size, so relation (4.9) requires that the ventricular pressure itself be independent of size. The same can, of course, be expected for the pressure in the arteries and veins, since these are determined by the ventricular pressure.

Thus, we see that both blood velocities and pressures in the arteries and veins at relative times in the pumping cycle are independent of animal size. Now, relative time in the pumping cycle is determined by the ratio of actual time to the period between heartbeats; or, equivalently, by the product of heart rate and time (ωt). We may accordingly express variations of blood velocity and pressure over the pumping cycle in the general mathematical forms as

$$V_1 = f_1(\omega t), \ P = f_2(\omega t) \qquad (4.10)$$

where f_1 and f_2 denote functions of dimensionless time which are the same for all mammals.

The above results are consistent with known scaling laws for blood velocities and pressures in the aorta of mammals. Independent of size, the peak and mean blood velocities in the aorta have values of about 100 cm/sec and 20 cm/sec, respectively, and values of the peak and mean arterial blood pressure of 120 mm Hg–160 mm Hg and 80 mm Hg–100 mm Hg, respectively (Kenner 1972). The variations of arterial pressure with time over the pumping cycle are also the same for animals of different size (Noordergraaf, Li, and Campbell, 1979). Some typical early blood pressure measurements on the mouse and dog are summarized in Table 4.5, along with accepted values for the human.

Blood Motions and Pressures in Capillaries

We may examine the scaling law for average blood velocity in the characteristic capillaries by observing that this velocity is equal to the average blood flow Q_b (or cardiac output) divided by the net cross-sectional flow area of the capillaries. Thus, with this area proportional to the product of the number of capillaries n_c and the square of their radius r_c, we have the capillary blood velocity v_c expressible as

$$V_c \;\alpha\; \frac{Q_b}{n_c\, r_c^{\,2}} \;\alpha\; W^{-1/24} \qquad (4.11)$$

where we have used the three-fourths power scaling relation for Q_b, as discussed above, and the relations (3.30) for n_c and r_c. We see that capillary blood, unlike arterial blood, increases in velocity with decreasing animal

TABLE 4.5.   Blood Pressures in Main Arteries of Mammals

Animal	Pressure(mm Hg)	
	Systolic	Diastolic
Mouse[a]	95–125	67–90
Dog[b]	126	89
Dog[c]	124	85
Human[d]	120	80

[a]Woodbury and Hamilton (1937), anesthetized mice; range of measurements from 9 mice.
[b]Gregg et al. (1937), anesthetized dog.
[c]Gregg et al. (1937), unanesthetized, well-trained dog.
[d]Generally accepted typical values for the human.

size, by animal weight raised to the minus one twenty-fourth power. This is an interesting result even though the value of the exponent is small. It means, for example, that the capillary blood in a mouse weighing 0.02 kg moves through the capillary with an average velocity that is about 1.6 times that for an elephant weighing 3000 kg.

In next considering capillary blood pressure, we may use Eq. (3.5) as a guide and express the drop in pressure ΔP_c at some distance X along a capillary as

$$\Delta P_c \ \alpha \ \frac{X}{L_c} \frac{L_c}{r_c^2} \ V_c \ \alpha \ \frac{X}{L_c} W^o \qquad (4.12)$$

Here we have used the above scaling law for V_c and the relations in Chapter 3 for L_c and r_c. We thus see that the pressure drop at any fixed relative distance X/L_c along the capillary will be independent of animal size. Since we know that the pressure at a capillary entrance is independent of animal size because of the previous consideration of arterial pressures, we may conclude that pressures in the capillaries are themselves independent of animal size as a consequence of the velocity dependence of relation (4.11).

Circulation Time

We have seen that the velocity of blood in the arteries and veins is predicted by the theory to be independent of animal size according to relation (4.7) and that the velocity of the blood in characteristic capillary vessels is dependent on animal size according to relation (4.11). The question then arises as to whether the time for complete circulation of an elemental volume of blood obeys any simple scaling law. To investigate this, we may consider the characteristic time needed for passage through the connecting tubes and characteristic capillaries of our system. Since time of passage is determined by the ratio of vessel length to blood velocity, the time through the connecting tubes is proportional to the ratio L_a/V_1. By the scaling law for L_a given in Chapter 3 and the above condition for size independent V_1, we see that this time is proportional to $W^{1/4}$. Similarly, the passage time through the capillary bed is proportional to the ratio L_c/V_c. Again, using the scaling relation for L_c from Chapter 3 and the above relation for V_c, we see that this time is also proportional to $W^{1/4}$. Assuming similar behavior for flow through the heart itself, we may thus express the total circulation time T_c as

$$T_c \ \alpha \ W^{1/4} \qquad (4.13)$$

that is, circulation time varies with animal weight raised to the one-fourth power.

Interestingly, this scaling relation has also been established by

Schmidt-Nielsen (1984) using simply the ratio of blood volume to average blood flow as the characteristic time measure. Such a ratio is of course dimensionally correct; however, it provides only an approximate numerical average value because it does not take into account the various routes than an elemental volume of blood can choose in making its circulation.

If we assume a typical average circulation time of 25 sec (McDonald 1968) for humans, we may use relation (4.13) to scale this value to other mammals. Results of this scaling are shown in Table 4.6.

We see from these results that the circulation time for the mouse is about one eighth that of the human, while that of the elephant is a little over twice the human value. Interestingly, if circulation times scaled directly with the characteristic length dimension of the body, that is, as animal weight to the one-third power, the circulation time for the mouse would then be only one-sixteenth that of the human, and that of the elephant would be more than three times the value for the human.

Prosser and Brown (1961) have collected some typical values for the rabbit (7.5 sec), the dog (16 sec), and man (23 sec). These are in good agreement with the values tabulated in Table 4.6.

We may note that the heart period, or time between heartbeats, is equal to the reciprocal of the heart rate, and that this also varies with animal weight raised to the one-fourth power. Thus, we see that the circulation time is directly proportional to the heart period for all mammals. It also follows from this observation that the time spent by an elemental volume of blood in passing through a capillary is directly proportional to the heart period under change of scale. The proportionality constants may be evaluated using data from the human. In particular, with a heart period of about 0.86 sec and a circulation time of about 25 sec, we find that circulation time is approximately equal to 29 times the heart period. Also, as noted in Chapter 2, the time required for an elemental volume of blood to pass through a capillary of the human is approximately 2.5 sec. Capillary-

TABLE 4.6   Average Circulation Times
for Various Animals as Scaled
from that of the Human

Animal	W (kg)	T_c (sec)
Mouse	0.02	3.2
Rabbit	2	10.2
Dog	20	18.2
Human	70	25.0
Horse	600	42.7
Elephant	3000	64.0

passage time for all mammals is therefore equal to about 2.9 times the heart period, or about one-tenth the circulation time.

4.3 SCALING LAWS DURING GROWTH AND AGING

We saw earlier that the scaling relations for heart rate and oxygen-consumption rate given by the present design theory are in precise agreement with the known scaling laws that apply to mammals of widely varying size. Interestingly, these scaling laws are also known to apply, at least approximately, to children during growth where their body weight increases from about 3–4 kg at birth to 60–70 kg on reaching maturity.

The application of the scaling law to measurements of the heartrate ω of children is illustrated in Fig. 4.5(a) using typical data of body weight, age, and heart rate collected by Altman and Dittmer (1974). The average reciprocal one-fourth power relation determined from the data is expressible as

$$\omega = 208 W_g^{-1/4} \qquad (4.14)$$

where W_g denotes body weight during growth.

Predictions from this equation are shown in Fig. 4.5(a) and are in good agreement with the measurements. The value of the coefficient in this equation is also in good agreement with the value of 230 appearing in the empirical law given in Chapter 1 and based on measurements on mature mammals ranging in size from mice to elephants.

Similar application of the scaling law for oxygen-consumption rate Q_o of children is illustrated in Fig. 4.5(b), again using data tabulated in the reference work of Altman and Dittmer (1974). The data are expressed here in units of oxygen volume rather than in equivalent heat units (4.8 kcal per liter of oxygen), as used in the reference. The average three-fourths power law determined from the data is expressible as

$$Q_o = 12.3 \ W_g^{3/4} \qquad (4.15)$$

The relation is seen to describe the data very well. It is likewise in good agreement with the empirical law given in Chapter 1, where a coefficient of 11.2 was determined from mature mammals over a wide range of body weights.

The above comparisons involving both heart rate and oxygen-consumption rate indicate, of course, a strong similarity between the design characteristics of the cardiovascular system at any stage of growth of a mammal and those existing for mature mammals. The design characteristics of the cardiovascular system of a mammal at any stage of its growth thus appear to be about the same as that of a smaller mature mammal

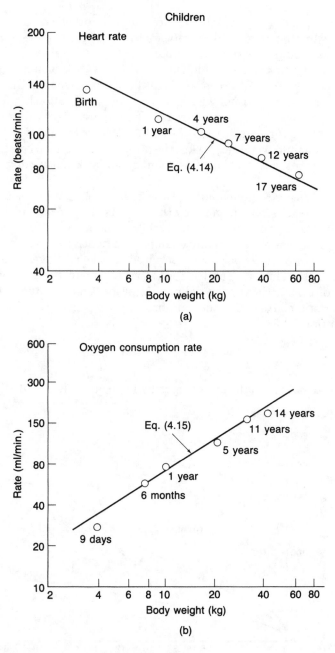

Fig 4.5 Comparison of (a) heart rate and (b) oxygen-consumption rate of children with predictions from Eqs. (4.14) and (4.15), respectively. Data source: Altman and Dittmer (1974).

having the same body weight. Or, to put it another way, intraspecies growth from offspring to adult appears to follow the same design theory as that describing interspecies comparisons.

We may provide some additional evidence for this similarity using measurements of aortic cross-sectional flow area given by Clark (1927). Thus, for newborn kittens with average weight of 0.126 kg, a value of 0.017 cm^2 is given for the aortic area. This compares well with the value of 0.018 cm^2 given for a mature rat of weight 0.180 kg. Similarly, for a one- to two-year old child weighing 10 kg, the aortic area is quoted as 1.0 cm^2, in good agreement with a value of 1.14 cm^2 given for a mature dog weighing 12 kg.

The indicated applicability of the present theory in describing the cardiovascular system during growth suggests that we may make further use of the theory to understand physiological changes that accompany growth. As an example, we may examine the relative increase in number and volume of average body cells serviced by the cardiovascular system during growth. Thus, if v_o and n_o denote, respectively, the cell volume and number of cells at birth, the volume v_s (proportional to the cube of the characteristic cell length) and number n_s during growth is expressible from relations (3.31) as

$$\frac{v_s}{v_o} = \left(\frac{W_g}{w_o}\right)^{3/8}, \frac{n_s}{n_o} = \left(\frac{W_g}{W_o}\right)^{5/8} \tag{4.16}$$

where W_g denotes weight at any specified age, as earlier, and W_o denotes

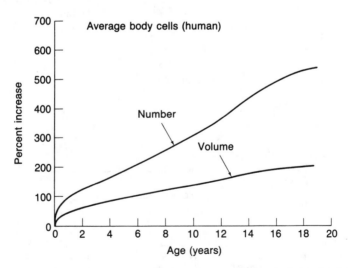

Fig 4.6 Theoretical percentage increase from birth of the number and individual volume of average body cells

weight at birth. The corresponding percentage increases are $100(v_s/v_o-1)$ and $100(n_s/n_o-1)$.

These percentages have been calculated using tabulated weight-age data for children (Altman and Dittmer 1974) and plotted in Fig. 4.6 against age. It can be seen that the increase in the number of cells contributes considerably more to the growth weight than does the increase in cell volume. At age 8, for example, growth has resulted in about a 250% increase in the number of cells since birth, and only about a 100% increase in cell volume. Similarly, at age 18, the number of cells has increased by more than 500% from birth, while the individual cell volume has increased by only about 200%.

As a further illustration of the use of the present theory to understand changes accompanying growth, we may consider the time needed for mammals to attain their mature body weight. For this purpose, we again concentrate attention on average body cells and assume, on the basis of dimensions, that increase in body weight per unit of time is proportional to the product of the mean rate of operation of the cells and the mean body weight during the growth period. We also assume, consistent with the present theory, that the mean rate of operation of the cells is proportional to the mean heart rate and that this is proportional to the mature body weight raised to the minus one-fourth power. In addition, the mean body weight during growth can be assumed proportional to the mature body weight when considering different mammals. With these conditions, the difference between mature body weight and birth weight, when divided by the time to reach maturity from birth, will then be proportional to the mature body weight raised to the three-fourths power. Finally, on neglecting birth weight in comparison with mature weight, or assuming it proportional under change in size, it then follows that the time for a mammal to reach maturity will be proportional simply to its mature body weight raised to the one-fourth power.

Predictions from this scaling relation are, in fact, in good agreement with the maturing times observed for various mammals. To illustrate this, we first note the well-known fact that a small mature dog weighing, say, 15 kg generally reached maturity about 2.5 years after birth. With these values, we can therefore use the above scaling law to predict the time T_m needed for other mammals to reach their mature body weight. Thus, for example, the maturing time for a horse weighing 600 kg when fully grown may be predicted from the equation

$$\frac{T_m}{2.5} = \left(\frac{600}{15}\right)^{1/4}$$

Solving for T_m, we find a value of 6.3 years. Such a value is, in fact, in good agreement with ordinary experience with horses. For an additional example, we may consider a small mouse weighing 0.025 kg when grown. Making a similar calculation, we then find the time needed to reach maturity to be only about 0.5 years. This result is also in good agreement with experience.

Equally valid predictions from the one-fourth scaling law for the maturing time of a number of other common mammals are indicated by empirical studies of Gunther and Guerra (1955). Interestingly, however, the case of the human presents an exception. In particular, for a human weighing 70 kg when grown, the above scaling procedure gives the time needed to reach maturity as only 3.7 years. This value is, of course, low by a factor of about four. The interpretation would appear to be that average body cells of humans require about four times as many cycles of operation as other common mammals in order to increase their body weight any fixed percentage during growth.

One final related observation is of interest and worthy of comment. This concerns the time required for old age and death of mammals. We assume, in particular, that aging of a mammal at any age after it has matured is proportional to the product of the rate of operation of average body cells and the elapsed time since maturity. In this case, the difference between the lifetime and the maturing time of a mammal is predicted to vary also with mature body weight raised to the one-fourth power. Moreover, because the maturing time since birth follows a similar scaling law, the total lifetime of a mammal is predicted to scale in this manner.

This relation is in good general agreement with the lifetimes of various mammals. To see this, we may consider again the small dog of 15-kg weight and note that its lifetime can be expected to be about 15 years. Using these values, we then find the lifetime of the 600-kg horse considered earlier to be predicted by the equation

$$\frac{T_L}{15} = \left(\frac{600}{15}\right)^{1/4}$$

where T_L denotes lifetime. On solving this relation, we find a value of T_L of 38 years for the horse which, like that for its maturing time, is in good agreement with experience. For the small mouse of 0.025-kg weight considered earlier, the lifetime is likewise calculated to be 3 years, a value also in good agreement with typical experience.

As in the case of maturing times, we find, however, that the human presents an exception to this scaling law. For the human weighing 70 kg, we find with the above scaling procedure a value for lifetime of only 22 years. This, of course, is low by about the same factor of four as found for the maturing time. The suggested conclusion from this and the previous result is that both further relative growth at any stage of development and further relative aging at any mature age involves for the human about four times as many cycles of operation of individual average cells as for other animals.

4.4 OXYGEN PARTIAL PRESSURES IN BLOOD

We have already seen in Chapter 3 that the present theory provides an accurate scaling law for the net oxygen-consumption rate of mammals. This rate represents, of course, the summation of the oxygen uptake of all the tissue cells of the body. If we examine the arguments in Chapter 3 leading to the governing scaling relation, we will find the implicit assumption that the difference in partial pressures (or, more generally, the difference in concentration of necessary substances) on opposite sides of the surface of an average body cell is independent of animal size. We may also see this directly by considering the average tissue cell illustrated in Fig. 4.7. As indicated, the oxygen partial pressures immediately outside and inside the cell are denoted by Po''_2 and Po'''_2, respectively. From the gaseous diffusion law, we know that the rate of oxygen movement into the cell is directly proportional to the product of the difference in these partial pressures and the membrane surface area and is inversely proportional to the membrane thickness [see Eq. (2.1) for an analogous statement for a capillary vessel]. In terms of the characteristic length ℓ_s of the cell, we accordingly have

$$\frac{Q_o}{n_s} \, \alpha \, \frac{(Po''_2 - Po'''_2)\ell_s^2}{\ell_s}$$

which may be rearranged and written as

$$Po''_2 - Po'''_2 \, \alpha \, \frac{Q_o}{n_s \ell_s} \tag{4.17}$$

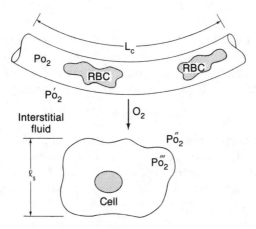

Fig. 4.7 Designation of oxygen partial pressures at a capillary and adjacent tissue cell

where Q_o/n_s represents the oxygen consumption rate of the average cell. Substituting into this relation the scaling laws for Q_o, n_s, and ℓ_s, we find

$$Po_2'' - Po_2''' \; \alpha \; W^o \tag{4.18}$$

that is, that the difference in oxygen partial pressures on opposite sides of an average body cell is required (or assumed) to be independent of animal size.

We may take the above restriction one step further by noting that, for similar cell processes among mammals, the partial pressure in the interior of the cell must be constant, independent of size. In this case, we then see that relation (4.18) requires that the partial pressure immediately outside the cell also be independent of animal size.

But what about the oxygen partial pressure in the blood itself? We observed in Chapter 2 that the oxygen partial pressure in the blood for any fixed level of saturation is higher for small mammals than for larger ones. We may now examine the matter in greater detail using the present theory. Consistent with the physiological process and in accordance with the arguments given in Chapter 2, we may consider that the net rate of diffusion of oxygen from the blood in a capillary to the interstitial fluid immediately outside it varies directly with surface area of the capillary and inversely with its wall thickness. Under change of scale, the wall thickness may be assumed to vary in the same way as the capillary radius. The driving force for the diffusion may also be taken equal to the difference in the average oxygen partial pressures, averaged over the capillary length, that exists between the blood and the adjacent interstitial fluid. Thus, similar to Eq. (2.2) we have

$$Po_2 - Po_2' \; \alpha \; \frac{Q_o}{n_c L_c} \tag{4.19}$$

where, as shown in Fig. 4.7, Po_2 and Po_2' denote average oxygen partial pressures in the blood and in the adjacent interstitial fluid, respectively. If we now substitute the scaling laws from the theory for Q_o, n_c and L_c, we find that the difference in pressures must vary according to the relation

$$Po_2 - Po_2' \; \alpha \; W^{-1/12} \tag{4.20}$$

Assuming local similarity in the process, such that the ratio of the two partial pressures must be constant, independent of size, we see that the partial pressure of each must then scale in the same manner as the difference. Thus, the scaling law for the oxygen partial pressure Po_2 in the blood must scale according to the relation

$$Po_2 \; \alpha \; W^{-1/12} \tag{4.21}$$

that is, as a reciprocal one-twelfth power law with animal weight.

Measurements of the partial pressure of oxygen in the blood of animals do, in fact, show an inverse dependence on animal weight like that predicted by relation (4.21). This was illustrated in general terms in Fig. 2.3 of Chapter 2. It is shown more explicitly in Fig. 4.8(a) where the saturation level of oxygen in the blood is plotted against the oxygen partial pressure for blood samples taken from the horse, the rabbit, and the mouse. These oxygen-dissociation curves are based on data obtained by Schmidt-Nielsen and Larimer (1958) and are typical of those found by them for a variety of mammals. For any fixed saturation level, it can be seen that the partial pressure, or unloading pressure as it is sometimes called, increases with decreasing size, being least for the horse and greatest for the mouse.

The predictions of relation (4.21) may be demonstrated more directly by choosing partial-pressure values associated with, say, a 75% saturation level as being typical of the average partial pressures in the capillaries of mammals. These have been taken from the experimental curves in Fig. 4.8(a) and plotted against the indicated animal weight in Fig. 4.8(b), using logarithmic-scale axes in the usual manner. Also shown are the corresponding values for the scaling relation (4.21), expressed in the form

$$Po_2 = 58 \, W^{-1/12} \qquad (4.22)$$

where the proportional coefficient has been determined from the average of the three measurements. It can be seen that the agreement with the reciprocal one-twelfth scaling relation is, indeed, very good.

As a further check on the validity of Eq. (4.22), we may apply it to the blood of man, weighing, say, 70 kg. In this case, the formula gives a value $Po_2 = 40.7$ mm Hg, in excellent agreement with the generally accepted value of 40 mm Hg for a 75% saturation level (see Guyton 1971; also Fig 2.3 of Chapter 2).

It should be noted that Schmidt-Nielsen and Larimer (1958) determined from their full set of measurements that the partial pressure at 50% saturation level varied with animal weight raised to the –0.054 power. This empirical value is, of course, somewhat less than the theoretical value of –1/12 (or –0.083) predicted by relation (4.21) for the average partial pressure and used in writing Eq. (4.22). The difference is, however, probably not significant considering the overall small dependence on animal weight and the variability of individual measurements. This may be illustrated by considering the various data in the manner shown in Fig. 4.9. Here the measured partial-pressure values, at 75% saturation level, have been multiplied by the associated animal weight raised to the one-twelfth power and the product then plotted against animal weight. If the data followed Eq. (4.22) exactly, they would all fall on the horizontal line indicated on the graph. There is, of course, some scatter about this line,

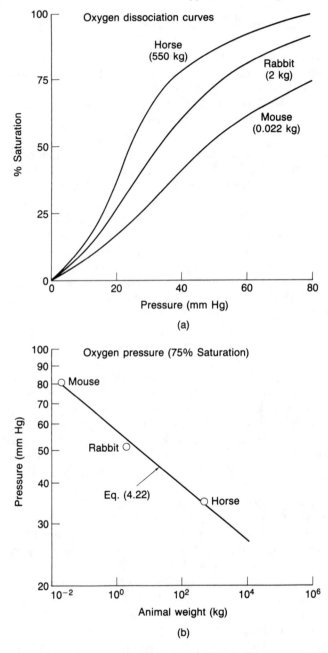

Fig. 4.8 (a) Oxygen-dissociation curves for the blood of the horse, rabbit, and mouse; (b) comparison of measurements of oxygen partial pressure with the reciprocal one-twelfth power law of Eq. (4.22). Data source: Schmidt-Nielsen and Larimer (1958).

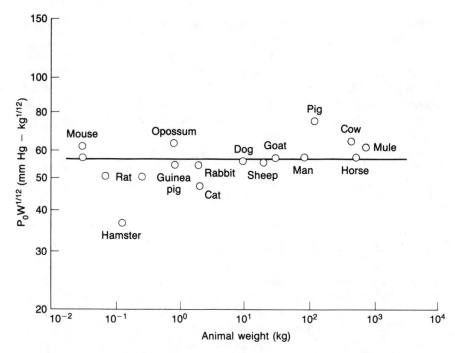

Fig. 4.9 Measurements of oxygen partial pressure by Schmidt-Nielsen and Larimer (1958) at 75% saturation demonstrating general agreement with the reciprocal one-twelfth scaling law from theory

but the general agreement can be seen to be very good with the exception of the values for the pig and hamster. These latter two values deviate in such a way as to indicate a slight increase in the plotted values with animal weight. However, such a trend is not supported by the remaining data. If all the data are considered, a best-fit analysis gives

$$Po_2 W^{1/12} \; \alpha \; W^{.02} \; \text{or} \; Po_2 \; \alpha \; W^{-.06}$$

in close agreement with the empirical relation found by Schmidt-Nielsen and Larimer (1958). On the other hand, if the data for the pig and hamster are put aside as being inconsistent with the remainder, we then find the best fit relation

$$Po_2 \; W^{1/12} \; \alpha \; W^{-.008} \; \text{or} \; Po_2 \; \alpha W^{-.09}$$

in good agreement with that required by the present theory for the average oxygen partial pressure.

In summary, we have seen from the results of the present theory that the average oxygen partial pressure along a capillary must increase with decreasing mammal size if the rate of oxygen transfer from a capillary is to equal the rate of consumption of oxygen by the cells serviced by it. We have also seen that conditions necessary for meeting this requirement are contained within the blood itself, with higher partial pressures associated with the blood of smaller animals for any fixed saturation level. The physical or chemical aspect of the blood responsible for this size-dependent variation is, unfortunately, not presently understood. However, its origin makes no difference to the present theory, which suggests that the increased average oxygen partial pressure in the capillary blood of smaller mammals gives rise to a resulting increase immediately outside the capillary in the interstitial fluid. This increase accordingly facilitates the diffusion of oxygen onward to the cells where scale-invariant oxygen partial pressures exist.

4.5 CAPILLARY DENSITY IN MUSCLE TISSUE

Consider next the scaling relation for the density of capillary vessels in muscle tissue where the vessels are approximately parallel to one another and run between the muscle fibers. We shall denote the average spacing between the vessels by d_c. From dimensional reasoning, the number per unit of cross-sectional area perpendicular to their length, that is, the capillary density, must be inversely proportional to the square of this spacing. For a simple design condition, we may assume similarity in the muscle cross section such that the spacing of the capillaries is proportional to their radius r_c. We then have, with the help of the first of relations (3.30), a possible scaling relation for capillary density in the form

$$\gamma \; \alpha \; W^{-1/6} \tag{4.23}$$

This relation predicts that the capillary density in muscle tissue will be greater in smaller animals than in larger ones. Such a prediction is in general agreement with limited measurements by Krogh (1919) who found, for example, the capillary density in guinea pig muscle to be more than twice that in horse muscle.

More recently, Schmidt-Nielsen and Pennycuik (1961) performed a detailed study of capillary density in the masseter (jaw) and gastrocnemius (leg) muscles of mammals ranging in size from bats to cows. The results from the gastrocnemius muscle are difficult to evaluate because of variable (non-similar) proportions of different types of muscle fibers among the various species. However, in the case of the masseter muscle, the structure is relatively uniform, thus allowing reasonable comparison between species.

The data for the masseter muscle are tabulated in Table 4.7, together with corresponding animal weights. For comparison with the theoretical

TABLE 4.7.   Measurements of Capillary Density for Masseter Muscle of
Various Animals by Schmidt-Nielsen and Pennycuik (1961)

Animal	W (kg)	γ (no./mm^2)	$\gamma W^{1/6}$ (mm$^{-2} - kg^{1/6}$)
Bat	0.009	3207	1462
Mouse	0.027	2272	1244
Rat	0.255	1220	971
Guinea pig	0.889	1704	1670
Cat	1.48	760	811
Rabbit	2.70	1049	1238
Dog	19	830	1356
Sheep	23	1427	2406
Pig	75	695	1427
Cow	454	956	2650

relation (4.23), values of the product $\gamma W^{1/6}$ have also been tabulated in
Table 4.7. It can be seen that the individual values are relatively consistent with one another, with the exception of those for the sheep and cow.
Excluding these two, the average is found to be 1270. In contrast, the values for the sheep and cow are seen to be about twice this average value.

These larger values are further seen to be a direct result of the larger
capillary densities found in the masseter muscle. Since both sheep and
cows are ruminating (cud-chewing) animals, these values may simply indicate that their masseter muscles are more highly developed than those of
the others listed in Table 4.7 (Schmidt-Nielsen 1984). In any case, putting
these measurements aside as being inconsistent with the others, we may
use the above average value and write the theoretical scaling relation
(4.23) in equation form as

$$\gamma = 1270 \ W^{-1/6} \tag{4.24}$$

This equation has been graphed in Fig. 4.10 using logarithmic-scale
axes for animal weight and capillary density. Also shown are the measurements of Table 4.7. The correlation is seen to be very good. This agreement is confirmed using a best-fit analysis of the data of Table 4.7 when
applied to a general power-law expression. Excluding the data from the
sheep and cow, we find a power-law exponent of -0.163 which differs from
the theoretical negative one-sixth relation by only about 2%. Thus, for the
masseter muscle, this initial design assumption concerning the spacing of
the capillaries appears correct; that is, that the spacing is proportional to
the capillary radius.

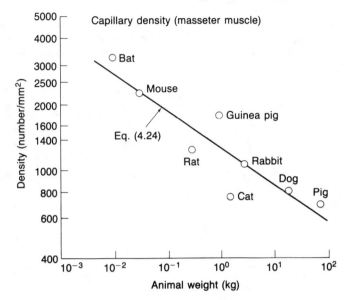

Fig. 4.10 Comparison of measurements of capillary density of masseter muscle given in Table 4.7 with reciprocal one-sixth power law of Eq. (4.24)

REFERENCES

ALTMAN, P. L., and D. S. DITTMER, editors. 1974. *Biology Data Book* Vol. III. Washington, DC: Federation of American Societies for Experimental Biology.

CLARK, A. J. 1927. *Comparative Physiology of the Heart*. Cambridge: Cambridge Unidersity Press.

GREGG, D. E., R. W. ECKSTEIN, and M. H. FINEBERG. 1937. Pressure pulses and blood pressure values in unanesthetized dogs. *Am. J. Physiol.* 118: 399–410.

GUNTHER, B. and E. GUERRA. 1955. Biological similarities. *Acta Physiol. latinoamer.* 5: 169–86.

GUYTON, A. C. 1971. *Textbook of Medical Physiology*. Philadelphia: W. B. Saunders.

HOLT, J. P., E. A. RHODE, and H. KINES. 1968. Ventricular volumes and body weight in mammals. *Am. J. Physiol.* 215(3): 704–15.

KENNER, T. 1972. Flow and pressures in the arteries. In *Biomechanics—Its Foundations and Objectives*. Y. C. Fung, N. Perrone, M. Anliker, eds., pp. 381–434. Englewood Cliffs: Prentice-Hall.

KROGH, A. 1919. The number and distribution of capillaries in muscles with calculations of the oxygen pressure head necessary for supplying the tissue. *J. Physiol* 52: 409–15.

McDONALD, D. A. 1968. Elementary hydrodynamics of the circulation. *In Princi-*

ples of Human Physiology, H. Davson and M. G. Eggleton, eds., pp. 222–51. Philadelphia: Lea and Febiger.

NOORDERGRAAF, A., J. K. J. LI, and K. B. CAMPBELL. 1979. Mammalian hemodynamics: a new similarity principle. *J. Theor. Biol.* 79: 485–89.

PROSSER, C. L., and F. A. BROWN, Jr. 1961. *Comparative Animal Physiology.* Philadelphia: W. B. Saunders.

SCHMIDT-NIELSEN, K. 1984. *Scaling: Why is Animal Size So Important?* Cambridge: Cambridge University Press.

SCHMIDT-NIELSEN, K., and J. L. LARIMER. 1958. Oxygen dissociation curves of mammalian blood in relation to body size. *Am. J. Physiol.* 195(2): 424–28.

SCHMIDT-NIELSEN, K., and P. PENNYCUIK. 1961. Capillary density in mammals in relation to body size and oxygen consumption. *Am. J. Physiol.* 200(4): 746–50.

WOODBURY, R. A., and W. F. HAMILTON. 1937. Blood pressure studies in small animals. *Am. J. Physiol.* 119(4): 663–74.

5

Special Consideration
of Individual Organs

The agreements between theory and measurement found in the previous chapter encourage further application of the theory in understanding the basic design of the cardiovascular system. In the present chapter, we therefore give special consideration to individual organs of the body.

5.1 OXYGEN UPTAKE OF ORGANS

According to concepts introduced in Chapter 3, the net oxygen-consumption rate of mammals is proportional under size change to the product of the number n_s and characteristic length ℓ_s of average body cells. Theoretical considerations of Chapter 3 also require that n_s and ℓ_s vary with animal weight raised to the five-eighths and one-eighth powers, respectively, so that this product leads directly to the three-fourths power law known to apply to mammals of widely varying size.

We now consider individual organs of the body rather than the body as a whole. In this task, we are confronted with the interesting fact that the weights of several of the major organs of the body do not vary directly with animal weight (Brody 1945; Adolph 1949). In discussing the oxygen consumption of the individual organs within the context of the present theory, we must therefore account for this additional feature.

With the above in mind, we introduce the concept of an average organ cell, defined such that the product of the number of these cells and their individual volume is proportional to the organ weight. We denote the characteristic length of the average organ cell by $\tilde{\ell}_s$ and the number of such cells in the organ by $\tilde{n}_s$. We also let $\tilde{\omega}$ denote the characteristic rate of operation of the cells.

Now, for similar cell processes among mammals of different size, the mass of a substance diffusing into, or out of, a cell during a cycle of operation must be proportional to the cell volume. This same observation was noted in Chapter 3 in our consideration of average body cells. Hence, by the same dimensionless arguments for diffusion (or equivalent processes) that were used in establishing the design Eq. (3.28), we have the relation

$$\tilde{\ell}_s \; \alpha \; \tilde{\omega}^{\,-1/2} \tag{5.1}$$

that is, that the characteristic cell length must be inversely proportional to the square root of the characteristic rate of operation of the cell.

Consistent with the above reasoning, as well as that used in Chapter 3, we note that the total oxygen consumption of an organ during a cycle of operation will be proportional to the product $\tilde{n}_s\tilde{\ell}_s^3$, and its rate of consumption will be proportional to the product $\tilde{\omega}\tilde{n}_s\tilde{\ell}_s^3$. Using relation (5.1), the total oxygen-consumption rate $\tilde{Q}o$ of an organ can therefore be expressed as

$$\tilde{Q}_o \; \alpha \; \tilde{n}_s\tilde{\ell}_s \tag{5.2}$$

In developing this relation, it is implicitly assumed, as in the analogous development in Chapter 3, that the oxygen-partial pressures immediately inside and outside the cell are invariant under change of scale; that is, they do not change when considering animals of different size.

We next assume similarity in organ functions such that each organ of the body operates at an overall organ rate (as distinct from the cell rate) that is proportional to the rate of operation of the body as a whole. Since oxygen uptake provides a measure of these rates, each organ must, under this assumption, consume oxygen at a rate proportional to the net consumption rate of the body; that is, proportional to animal weight to the three-fourths power. The scaling law for oxygen-consumption rate of an organ is accordingly expressible as

$$\tilde{Q}_o \; \alpha \; \tilde{n}_s\tilde{\ell}_s \; \alpha \; W^{3/4} \tag{5.3}$$

The condition that the oxygen-consumption rate of an organ obeys the same scaling relation as that for the entire body requires that the vascular scaling relations of each organ also follow the relations developed in Chapter 3 for the body as a whole. Otherwise, an imbalance between oxygen delivery and uptake would exist in the organs of mammals of different size.

We may examine the validity of the scaling aspect of relation (5.3) using measurements reported by Martin and Fuhrman (1955) for the mouse and the dog. In this work, the oxygen-consumption rate of each major organ was determined by multiplying the measured tissue respiration rate per unit weight by the measured organ weight. An abbreviated version of their results is given in Table 5.1. Here, as opposed to the original tabulations, the values for oxygen uptake of the skeletal muscle of the dog and mouse have been determined by forming the difference between the net oxygen-consumption rates, given by Eq. (1.6), and the subtotals of the measurements from all other organs and systems of each animal.

For examination of relation (5.3) and the underlying similarity assumption, values of the oxygen-consumption rate of the organs from the dog have been scaled to corresponding values for the mouse using the assumed three-fourths scaling law in the form

$$\bar{Q}_o \text{ (mouse)} = 0.0056 \, \bar{Q}_o \text{ (dog)}$$

In this equation, the numerical factor of 0.0056 is the value of the ratio of mouse weight (0.02 kg) to dog weight (20 kg) raised to the three-fourths power. Results from these calculations are shown in Fig. 5.1 where the scaled values are plotted against the measured values for the mouse organs. Also shown in the figure is a straight line with slope of 1.0. If the agreement of the scaled values with the measured values were exact, all data points would lie along this line. Although the values show some scat-

TABLE 5.1   Weights ($\bar{W}$) and Oxygen-Consumption Rates($\bar{Q}_o$) of Organs or Organ Systems from Tissue-Respiration Studies of Martin and Fuhrman (1955)

Organ or System	Dog (20 kg)		Mouse (0.02 kg)	
	$\bar{W}$ (g)	$\bar{Q}_o$ ml/hr	$\bar{W}$ (g)	$\bar{Q}_o$ ml/hr
Liver	438	898	1.24	4.13
Kidneys	90	222	0.36	1.81
Skin	2568	436	2.85	1.37
Brain	114	156	0.36	1.11
Lungs	154	75	0.14	0.21
Digestive tract	694	601	1.49	3.44
Fat	778	202	2.07	0.89
Remainder (less muscle)	7020	365	4.56	2.61
Subtotal	11856	2955	13.07	15.57
Muscle	8144	3395*	6.93	20.13*
TOTAL	20000	6350**	20	35.70**

*Calculated as the difference between total and subtotal.
**Calculated from Eq. (1.6).

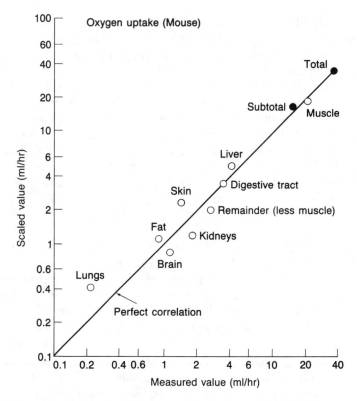

Fig. 5.1 Comparison of scaled values of the oxygen uptake of various organs of the mouse with measured values. The scaled values were derived from measurements for the dog using the three-fourths scaling law of relation (5.3). Basic data used are those given in Table 5.1.

ter about the line, there is no discernible trend and the results provide general support for the scaling relation (5.3).

We now consider the scaling relation for organ weight. From the definition of the average organ cell and the fact that the volume of the cell is proportional to the cube of its characteristic length, we may express the organ weight $\tilde{W}$ as

$$\tilde{W} \; \alpha \; \bar{n}_s \bar{\ell}_s^3 \tag{5.4}$$

Thus, the organ weight depends only on the characteristic length and number of average organ cells through this product. We may regard the characteristic length of the cells to be fixed by their rate of operation according to relation (5.1). Similarly, the number of cells is fixed by the overall rate of operation of the organ, as represented by relation (5.3). The scaling law for organ weight is therefore determined by these two rate processes. Clearly, the resulting scaling law for the weight of an organ does not have to be a simple proportional relation with body weight W. In fact, it is easily seen that this will only be the case when the rate of operation of the cells is proportional to the heart rate; for, in this case, relation (5.1) requires that $\tilde{\ell}_s \propto W^{1/8}$, relations (5.3) require $\tilde{n}_s \propto W^{5/8}$, and relation (5.4) then requires $\tilde{W} \propto W$. For other rates of operation, different scaling laws for organ weight must result.

It should be noted that the above special-case scaling laws for organ cells are the same as those for average body cells found in Chapter 3. Thus, organs whose weights are directly proportional to body weight may be considered to be composed of average body cells, with their rate of operation proportional to heart rate. From our earlier discussion, we know the heart is such an organ. It should come as no suprise that the rate of operation of its cells is proportional to heart rate. Such a condition was, in fact, referred to in Chapter 3.

In contrast to predicting the scaling law for organ weight from those of its cells, we may use measurements to describe the scaling law for the organ weight and infer the scaling relations for the cells. To carry out this operation, we may represent the organ weight and the characteristic length and number of average organ cells as power laws of body weight in the form

$$\tilde{W} \propto W^{\tilde{a}}, \quad \tilde{\ell}_s \propto W^{\tilde{b}}, \quad \tilde{n}_s \propto W^{\tilde{c}} \tag{5.5}$$

where the exponents $\tilde{a}$, $\tilde{b}$ and $\tilde{c}$ are required by relations (5.3) and (5.4) to satisfy the equations

$$\tilde{b} = \frac{\tilde{a}}{2} - \frac{3}{8}, \quad \tilde{c} = \frac{9}{8} - \frac{\tilde{a}}{2} \tag{5.6}$$

The values of the above exponents have been worked out for several major organs of the body using the organ-weight data given in Table 5.1.

TABLE 5.2.   Scaling Exponents for Selected Major Organs

Organ	$\tilde{W} \propto W^{\tilde{a}}$	$\tilde{\ell}_s \propto W^{\tilde{b}}$	$\tilde{n}_s \propto W^{\tilde{c}}$	$\tilde{\omega} \propto W^{\tilde{d}}$
	$\tilde{a}^*$	$\tilde{b}$	$\tilde{c}$	$\tilde{d}$
Liver	0.85	0.05	0.70	−.10
Kidneys	0.80	0.03	0.72	−.06
Skin	0.99	0.12	0.63	−.24
Lungs	1.01	0.13	0.62	−.26
Muscle	1.02	0.14	0.61	−.28

*From data of Table 5.1.

The results are listed in Table 5.2. Also included are values of the associated exponent $\tilde{d}$ for the scaling law for cell rate, as described from relation (5.1) by

$$\tilde{\omega} \; \alpha \; W^{\tilde{d}}, \; \tilde{d} \; = \; -2\tilde{b} \qquad (5.7)$$

It can be seen that the weights of the skin, lungs, and muscle scale directly with animal weight and that the scaling laws for the characteristic length and number of average cells of these organs are essentially the same as those for the average body cells determined in Chapter 3, namely, as animal weight raised to the one-eighth and five-eighths powers, respectively. Cell rates are also essentially proportional to heart rate. In contrast, the weights of the liver and kidneys are seen to scale with animal weight raised to powers of 0.85 and 0.80, respectively. The corresponding powers for the characteristic cell lengths are 0.05 and 0.03 and those for the numbers of cells are 0.70 and 0.72. We see that the characteristic lengths of these average cells and their rates of operation do not vary with size nearly as much as the average body cells.

It is worth noting that the scaling-law exponent for liver weight given in Table 5.2 is in good agreement with values found by others even though it is based on measurements from only two species. Thus, Brody (1945) reported for a wide range of mammals a value of 0.87, Munro and Downie (1964) gave a value of 0.84, and Prothero (1982) gave a value of 0.89, all in reasonable agreement with the value of 0.85 found here. It is also of interest to note that Smith (1955) found from measurements that the total number of mitochondria (the respiration units of a cell) in the liver scaled with animal weight to the 0.72 power. If we assume the number of mitochondria per liver cell to be constant, we then have the associated result that the number of cells in the liver scales with animal weight to the 0.72 power, a result in remarkably close agreement with the value of 0.70 given in Table 5.2. The scaling-law exponents for the weights of the kidneys and lungs listed in Table 4.6 are also in general agreement with other reported values. Brody (1945) for example, gave values of 0.85 and 0.99 for the kidneys and lungs, respectively, in comparison with the values of 0.80 and 1.01 used here.

The brain has not been included in the organs considered in Table 5.2 because it requires special attention. If we use the weight data given in Table 5.1, we find a scaling-law exponent for brain weight of 0.83. However, this value is significantly different from the value of 0.70 reported by Brody (1945) and based on a wide range of mammals. If the value of 0.70 is indeed representative, it would provide through Eqs. (5.6) a negative scaling-law exponent for characteristic cell length of –0.025, in contrast with positive values found for the other organs listed in Table 5.2. The negative sign means, of course, that the characteristic length of the brain cells decreases rather than increases with increasing animal size. The low exponent of 0.70 may, however, simply be the result of considering data from specialized animals, such as livestock, whose body weights may not be what their brains intended.

Table 5.3 lists values of brain weight and animal weight for relatively non-specialized mammals. The data have been taken from results collected by Brody (1945). However, data from livestock have been excluded, as have been data for humans whose brains obviously have a special place among mammals. If we apply best-fit methods to determine the appropriate scaling law for these data, we find that brain weight now scales with animal weight raised to the 0.75 power, rather than the 0.70 value.

Writing 0.75 as 3/4, the full scaling equation is expressible as

$$\tilde{W} = 7.6 \ W^{3/4} \tag{5.8}$$

This equation has been graphed in Fig. 5.2 using logarithmic-scales axes in our conventional manner. The measurements have also been plotted in this figure, being denoted by open circles. The data can be seen to follow the predictions of Eq. (5.8) in a consistent manner. Also shown in the figure is the value of brain weight for humans (1300 g). This is seen to be about an order of magnitude greater than that expected from Eq. (5.8). Finally, we see in the figure data for livestock (steer, cow, hog) from Brody's compilation. These data consistently fall below the values predicted by Eq. (5.8) as a result, perhaps, of their specialized use. If they are included with those in Table 5.3, the resulting best-fit scaling relations then requires brain weight to vary with animal weight to the 0.70 power, in agreement with Brody's value.

Returning to our original discussion of the scaling laws for brain cells, we see, with the three-fourths scaling law of Eq. (5.8), that the characteristic length of brain cells is now predicted to be independent of animal size and that the number of cells is predicted to vary with animal

---

**TABLE 5.3.   Measurements of Brain Weight $\tilde{W}$
for Animals of Weight W**

Animal	W (kg)	$\tilde{W}$ (g)
Rat	0.25	2
Guinea pig	0.80	4.7
Monkey	4.5	42
Dog	10	75
Sheep	52	106
Horse	600	670
Elephant	6650	5700

Data source: Brody (1945).

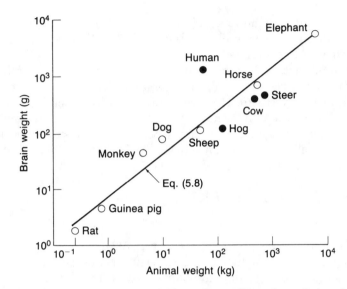

Fig. 5.2 Comparison of measured brain weights with best-fit predictions from Eq. (5.8). The data for the humans and the livestock were not included in the best-fit calculation. Data source: Brody (1945).

weight to the three-fourths power. We also have the interesting prediction that the rate of operation of the brain cells is independent of animal size.

On the basis of the results of this section, we may conclude that the vascular design and oxygen-consumption rates of the individual organs of the body obey the same scaling relations as those of the body as a whole. We may also conclude that the characteristic length and number of average tissue cells of certain organs do not follow the same scaling laws as those for average tissue cells of the entire body. The reason is the need for different rates of operation of the cells of these organs. This condition is also responsible for the non-proportional variation of the weights of these organs with animal weight.

5.2 BLOOD FLOW TO ORGANS

We saw in the previous section that the oxygen-consumption rates of individual organs obey the same type of scaling relation as that for the body as a whole. It was also noted that this condition implies, for oxygen balance, that the vascular system of the organs also scales in the same manner as that determined in Chapter 3 for the overall body. We may, in fact, show that the present design theory requires the average blood flow to any organ to be proportional to the cardiac output and, hence, to scale with animal weight in the same way that cardiac output does.

For this purpose, we may express the cardiac output Q_b in terms of the mean blood velocity V_1 in the aorta as

$$Q_b = V_1 \pi R^2$$

where R denotes the aortic radius. For a branching artery of radius $\tilde{R}$ supplying blood to an organ, as illustrated in Fig. 5.3, we may also express the associated average blood flow Q_b as

$$\tilde{Q}_b = \tilde{V}_1 \pi \tilde{R}^2$$

where $\tilde{V}_1$ denotes the average blood velocity in the artery. Forming the ratio $\tilde{Q}_b/Q_b$, we then have the relation

$$\frac{\tilde{Q}_b}{Q^b} = \frac{\tilde{V}_1}{V_1} \left(\frac{\tilde{R}}{R}\right)^2$$

Now, it was shown in Section 4.2 of the previous chapter that the velocities $\tilde{V}_1$ and V_2 are each independent of animal size so that their ratio is independent of size. The ratio of $\tilde{R}$ to R is likewise independent of size since each is required by relation (3.29) to vary with animal weight to the three-eighths power. Thus, we see that the ratio $\tilde{Q}_b/Q_b$ is independent of animal size or, equivalently, that the blood flow to an organ is proportional, under size change, to the cardiac output. Moreover, since cardiac output varies with animal weight to the three-fourths power, so also must organ blood flow.

We may illustrate the validity of this result by considering measurements of the blood flow to the liver. This organ is of particular interest since, as discussed in Chapter 2, it receives about 75% of its blood from the portal vein after this blood has first passed through the mesenteric system, the stomach, intestines, pancreas and spleen. The remainder (about 25%) of the blood supply is from the hepatic artery. In order for

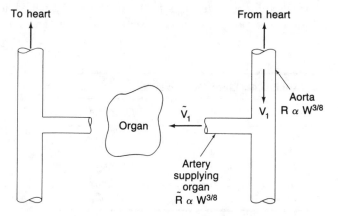

Fig. 5.3   Illustration of branching artery supplying blood to an organ

blood flow to the liver to vary with animal weight to the three-fourths power, it is necessary that the arteries supplying the net system and the liver as well as the portal veins all follow the three-eighths scaling law for the radius of connecting vessels.

An extensive number of measurements of the mean blood flow to the liver have been made for rats, rabbits, dogs, and humans. Prothero (1982) has collected and averaged these data from various sources to get representative values for each of the species. These are listed in Table 5.4.

Assuming the measurements scale with animal weight to the three-fourths power, we may determine the scaling equation from the data as

$$\tilde{Q}_b = 64.5 \ W^{3/4} \tag{5.9}$$

This equation has been graphed in Fig. 5.4 along with the individual measurements. It can be seen that the agreement is, indeed, very good. A best-fit analysis gives the equation

$$\tilde{Q}_b = 60.6 \ W^{0.78}$$

in good agreement with Eq. (5.9).

It should be mentioned that Prothero (1982) obtained a larger exponent of 0.91 in the flow equation by including additional less-extensive measurements from other species. It was noted, however, that such a large exponent was not likely to be representative of species variations, since it would require the blood flow to the liver in very large mammals to be an unreasonably large percentage of the total cardiac output.

We may combine Eq. (5.9) with the expression for cardiac output given by Eq. (1.8) to get the ratio

$$\frac{\tilde{Q}_b}{Q_b} = \frac{64.5W^{3/4}}{224W^{3/4}} = 0.29$$

which shows that the liver receives about 29% of the resting cardiac output. This value and corresponding ones for certain other organs, as deter-

TABLE 5.4   Blood Flow to Liver

Animal	W (kg)	$\tilde{Q}_b$ (ml/min.)
Rat	0.25	19.4
Rabbit	2.9	144
Dog	10	396
Dog	19	636
Human	67	1462

Data source: Prothero (1982).

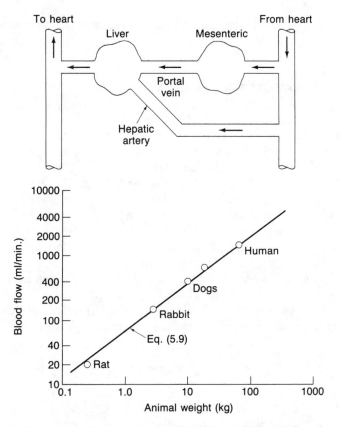

Fig. 5.4   Comparison of measurements of blood flow to liver with three-fourths scaling law in the form of Eq. (5.9). Data source: Prothero (1982).

TABLE 5.5	Fraction of Resting Cardiac Output to Various Organs.
Organ	$\bar{Q}_b/Q_b$
Liver	0.29
Kidney	0.21
Skin	0.07
Heart	0.05
Other	0.38

Data source: Davson and Eggleton, (1968).

mined from generally accepted values for humans, are tabulated in Table 5.5.

5.3 VASCULAR DESIGN OF KIDNEYS

As the blood circulates throughout the body, it absorbs water and waste products formed from cell metabolism. The circulation of the blood through the kidneys forms a special part of the systemic circulation where removal of this waste fluid takes place.

The basic unit in the kidney is the nephron. This fundamental unit extracts fluid from the blood and allows reabsorption of the valuable products and secretion of others such that waste water, or urine, is produced for discharge from the body. The two kidneys of mammals contain many thousands of these units, each operating continuously to produce urine that is stored in the bladder for periodic discharge.

Figure 5.5 illustrates the general features of a nephron. The basic operation is as follows: Blood flows into a capillary network referred to as the glomerulus from which fluid is filtered out of the blood through small pores in the capillary walls. The fluid is captured in the surrounding cap-

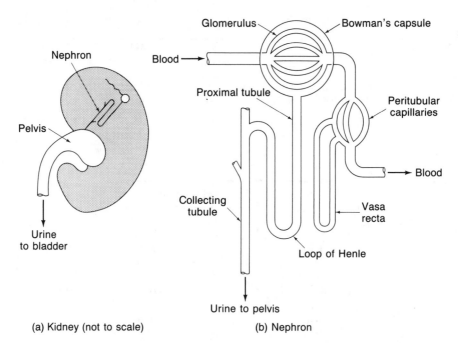

(a) Kidney (not to scale)   (b) Nephron

Fig. 5.5 Illustration of kidney design showing a fundamental nephron unit. The two kidneys of the human contain many thousands of these units.

sule (Bowman's capsule) and drained through tubules to the pelvis of the kidney, from which it empties into the bladder. Along its route, most of the water and varying amounts of its solutes are reabsorbed into the blood vessels surrounding these tubules (the vasa recta). The water and products not reabsorbed combine with secreted products to form the urine that flows onward to the bladder. The rate at which fluid is filtered out of the glomerulus in humans is typically about 125 ml/min., while the rate of urine production is typically only about 1 or 2 ml/min. Thus, a significant amount (98%–99%) of the fluid initially filtered out of the glomerulus is reabsorbed in the tubules.

Because of the two distinct sets of capillaries in the kidney, the glomerular and peritubular capillaries, the flow of the blood in the kidneys is through a series connection of these two capillary beds. Input to the glomerular capillaries is through the afferent arterioles and input to the peritubular capillaries is through the efferent arterioles. The series connection is illustrated in Fig. 5.6.

The average amount of blood flowing through both kidneys of the human is typically about 1200 ml/min., which amounts to about 21% of the total cardiac output. This percentage can also be expected to apply to other mammals, as noted in the previous section.

Nephra Number

In considering the design of kidneys of mammals of different size, the question naturally arises as to how the number of nephrons varies from one animal to the next. We may examine this question using the present theory. For this purpose, we make the initial design assumption that the number n' of glomerular capillaries in a single nephron is the same for all mammals. If N denotes the total number of nephrons in a mammalian kidney, the total number of glomerular capillaries is therefore

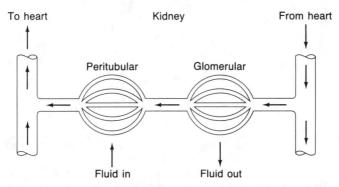

Fig. 5.6 Illustration of the series connection of the capillary beds in the kidney

expressible as Nn'. Since arguments in the previous two sections require the vascular scaling relations in the organs to be the same as for the body as a whole, this may further be regarded as being proportional to the characteristic number n_c of capillaries in the system. Since n' is assumed constant, we thus have the nephron-number scaling relation, in terms of animal weight W, expressible as

$$N \; \alpha \; n_c \; \alpha \; W^{5/8} \tag{5.10}$$

where the five-eighths scaling relation for n_c is that given by relation (3.30).

The above relation, requiring that the number of nephrons vary with animal weight to the $5/8 = 0.625$ power is, in fact, in remarkably close agreement with measurements made many years ago by Kunkel (1930). These data have been analyzed by Adolph (1949), who determined, using best-fit methods of analysis, that the number of nephrons varied with animal weight to the 0.62 power. Writing this exponent as 5/8 and expressing animal weight in kilograms, Adolph's formula may be written in terms of the number of nephrons per kidney as

$$N = 9.4 \times 10^4 \, W^{5/8} \tag{5.11}$$

The agreement of individual measurements with this relation is shown in Fig. 5.7, where typical data have been plotted on logarithmetric-

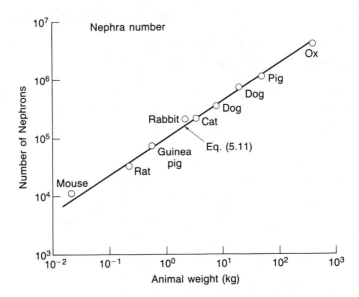

Fig. 5.7 Comparison of measurements of Nephra number with predictions from the theoretical five-eighths scaling law in the form of Eq. (5.11). Data source: Kunkel (1930).

scale axes along with predictions from Eq. (5.11). It can be seen that the correlation is, indeed, excellent. This result therefore provides additional support for the five-eighths power scaling relation for capillary number, as derived in Chapter 3. It also provides support for the initial design assumption of constant number of capillaries per nephron.

Urine Production

Continuing with our investigation of the design of the kidneys, we next consider the flow of glomerular fluid and urine production in mammals of different size. The walls of the capillaries of the glomerulus are known to be relatively porous in nature, with the diameter of the pores sufficiently small to prevent passage of blood cells and the proteins of the plasma, yet large enough to allow water and dissolved substances to pass through the walls. This process is referred to as ultrafiltration and is similar to that involved in the exchange of fluid with capillary blood in other parts of the body.

The driving force behind the fluid flow through the capillary walls is the blood pressure. This is, however, opposed to some extent by the osmotic pressure of the plasma proteins which tends to draw fluid into the capillary blood. It is also opposed by the pressure in the surrounding tissue. Thus, the net filtration pressure ΔP is expressible as

$$\Delta P = P_b - (P_{os} + P_t)$$

where P_b denotes the capillary blood pressure, P_{os} the osmotic pressure, and P_t the tissue pressure. The associated flow Q_f, or glomerular filtration rate, is determined from viscous fluid theory, assuming β pores per unit capillary surface area, and is expressible (Pappenheimer et al. 1951) as

$$Q_f = \frac{\beta \pi \, r_p^{\,4}}{8 \mu_f} \, \frac{A_g \Delta P}{h_c} \tag{5.12}$$

where r_p denotes the pore radius, A_g denotes the total surface area of all glomerula capillaries in both kidneys, h_c denotes the wall thickness of the capillaries, and μ_f denotes the viscosity coefficient of the fluid. An equation like this for a single capillary was also described earlier in Chapter 2.

Now, if we introduce the design assumption of similar material such that the pore density β and pore size r_p are constant, independent of animal size, we may write Eq. (5.12) as

$$Q_f = K_f \, \frac{A_g \Delta P}{h_c} \tag{5.13}$$

where K_f denotes the filtration constant defined by

$$K_f = \frac{\beta \pi r_p{}^4}{8\mu_f}$$

We observed earlier that the blood pressure in the capillaries and arteries is, according to the present theory, independent of mammal size. Assuming this is also the case for the osmotic and tissue pressures, we see from Eq. (5.13) that the glomerular flow is proportional to the total surface area of all glomerular capillaries and inversely proportional to their wall thickness. Thus, with n' denoting the number of glomerular capillaries in a single nephron and N denoting the total number of nephrons, as above, we have,

$$A_g = Nn'(2\pi r_c L_c)$$

where, as earlier, r_c and L_c denote the characteristic radius and length of the capillaries. Substituting this relation into Eq. (5.13), we find

$$Q_f = BNL_c \qquad (5.14)$$

where B is defined by

$$B = 2\pi K_f n'(r_c/h_c)\Delta P$$

From our earlier discussion, we may regard the number n' of glomerular capillaries in a nephron to be the same for all mammals. We also require here that the ratio of radius to wall thickness of the capillaries be the same for all mammals. The coefficient B in Eq. (5.14) is therefore constant, independent of mammal size. In this case, the glomerular flow is then seen to be proportional to the product of the total number of nephrons in the kidneys and the characteristic capillary length. But the total number of nephrons in each kidney (and hence in both) has been found proportional to the characteristic number of capillaries n_c according to relation (5.10). We accordingly have the design equation for the glomerular flow expressible as

$$Q_f \propto n_c L_c \propto W^{5/6} \qquad (5.15)$$

where the scaling relations (3.30) were used for n_c and L_c.

The applicability of this equation in predicting glomerular flow in mammals of different sizes may be examined using measurements of the removal, or clearance, of insulin from the blood plasma of animals. This substance is such that, when injected into the blood, it is filtered out by the glomerular flow and is neither absorbed nor secreted by the tubules. Such data have been collected from various sources by Adolph (1949) for the rat, rabbit, dog and human. The resulting glomerular flows for these animals are shown in Fig. 5.8 and contrasted with predictions from the five-sixth scaling relation:

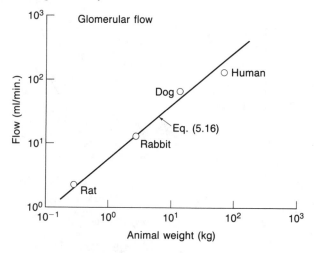

Fig. 5.8 Comparison of measurements of glomerular flow with predictions from the five-sixths scaling law in the form of Eq. (5.16). Data source: Adolph (1949).

$$Q_f = 5.6 \ W^{5/6} \tag{5.16}$$

as estimated from the data. The general agreement with these limited data is seen to be very good.

If we make the further design assumption that all reabsorptions and secretions in the nephron tubules are proportional to the glomerular flow, we have the scaling law for the urine flow Q_u similarly expressible as

$$Q_u \ \alpha \ Q_f \ \alpha \ W^{5/6} \tag{5.17}$$

This last relation, requiring urine output to vary with animal weight to the $5/6 = 0.83$ power is, like that of the nephra number, in remarkably close agreement with actual measurements on mammals. In particular, Adolph (1943, 1949), analyzed data from a variety of mammals ranging in size from mice to elephants and found with best-fit methods that urine output varied with animal weight raised to the 0.82 power. Writing 0.82 as $5/6$, and expressing animal weight in kilograms, Adolph's best-fit formula may be written as

$$Q_u = 0.031 \ W^{5/6} \tag{5.18}$$

where Q_u is expressed in units of ml/min. The agreement of this relation with the individual data is shown in Fig. 5.9, where excellent correlation is seen for all the measurements.

In connection with the above results, it is interesting to note that the

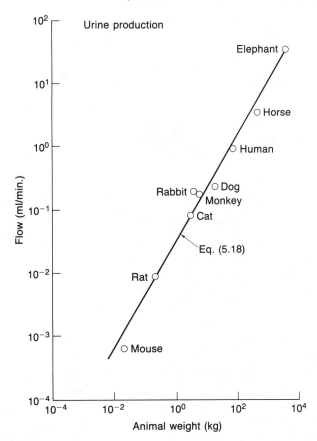

Fig. 5.9 Comparison of measurements of urine production with predictions from the five-sixths scaling law in the form of Eq. (5.18). Data source: Adolph (1943).

urine output per nephron is dependent on mammal size according to the relation

$$Q_u /N \ \alpha \ W^{5/24}$$

The urine output per nephron is therefore greater for large animals than for small ones, that of a 3000-kg elephant being about 11 times that of a 0.03-kg mouse. This fact was noted by Adolph (1949), but no explanation was offered for the difference. From the present theory, we see that the increased output of the nephra for the large animals can be attributed directly to the increased length of the glomerular capillaries in each nephron. From relations (5.14) and (5.17), the ratio Q_u /N is, in fact, seen to be directly proportional to the characteristic length L_c.

5.4 LUNG DESIGN

Blood returning to the heart from the systemic circulation is pumped by the right ventricle to the lungs for gas exchange with the outside environment. The function of the lungs is therefore to bring inspired air into contact with the circulating blood so that this gas exchange—in particular, the uptake of oxygen and the discharge of carbon dioxide—can readily take place.

The rate at which air is taken into the lungs, that is, the rate of breathing, of resting mammals is approximately one-fourth the heart rate (Stahl 1967). Thus, for man, the breathing, or respiratory, rate is approximately 70/4, or 18 breaths per minute. In contrast, for the mouse, it is approximately 600/4, or 150 breaths per minute.

Each of the two lungs of mammals has a main conducting tube for air, known as the bronchus. This branches from the windpipe, or trachea, and then branches further into smaller and smaller bronchial tubes which ultimately end in the terminal bronchioles. These feed air to very small air sacs called alveoli, which are connected through respiratory bronchioles and alveolar ducts, as indicated in Fig. 5.10.

In the human, there are several million alveoli in the two lungs, each having an average diameter of about 0.10 mm. The air in these sacs is separated from the interstitial space and pulmonary capillaries by a thin membrane so that gas exchange between blood and air is relatively unimpeded.

We noted earlier that the lungs have two separate blood flows: (1) the pulmonary flow, which takes blood from the right ventricle to the pulmonary capillaries for gas exchange and then returns it to the left ventricle for recirculation; and (2) the bronchial flow, which is that part of the systemic flow that takes blood to the lungs for nourishment of their tissues. An interesting aspect of this flow, also noted earlier, is that it returns directly to the left ventricle for further systemic circulation rather than to the right ventricle, as does other systemic venous blood.

Pulmonary Blood Vessels

The scaling relations for the blood vessels of the bronchial flow are, of course, those of the systemic circulation, as given by relations (3.29) and (3.30). It will be recalled that these relations were derived from the three general similarity conditions (3.21) and the two additional design relations (3.25) and (3.28), developed from considerations of cardiac excitation and cardiac (or average systemic) cellular dynamics, respectively. In the case of the blood vessels of the pulmonary circulation, these relations can still be expected to apply. Alternatively, if we consider the relation (3.28) to apply strictly to average systemic conditions, we may replace it

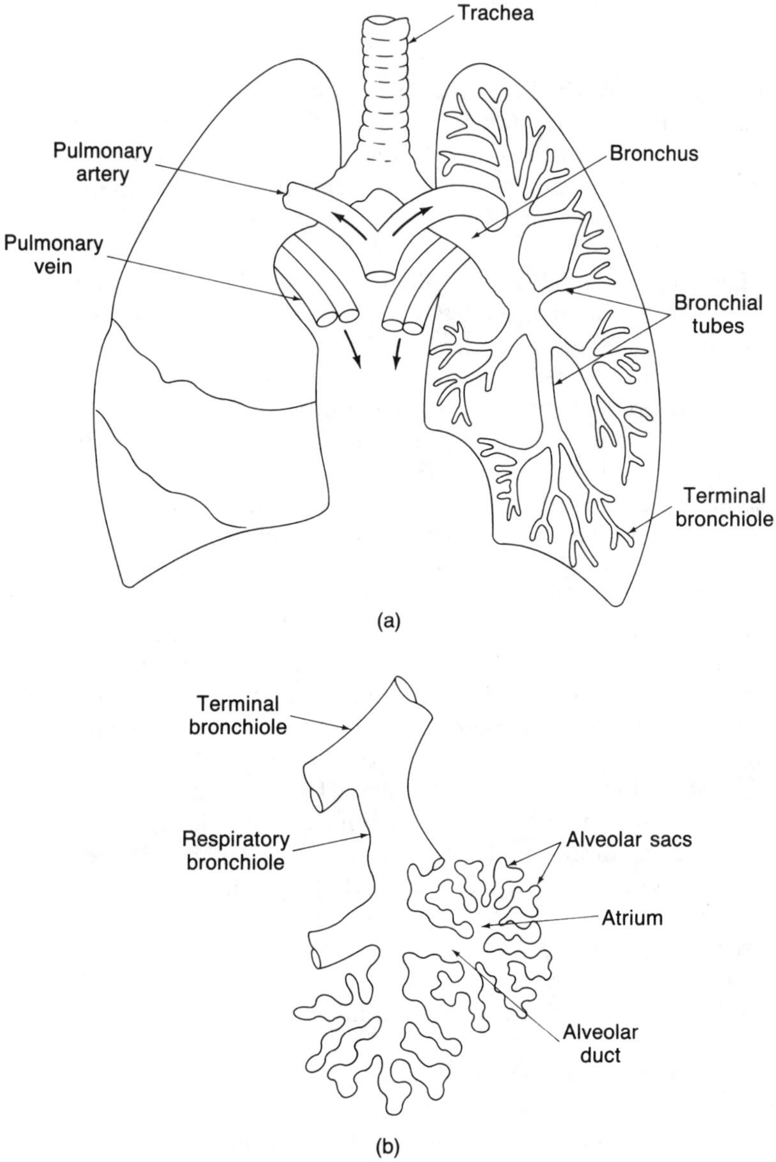

(a)

(b)

Fig. 5.10.   Illustration of (a) lung design and (b) design of alveolar respiratory
unit branching from terminal bronchiole

by the condition that the net oxygen uptake of the pulmonary capillaries
must equal the oxygen consumption rate of the body and, hence, must be
proportional to body weight to the three-fourths power. This relation may
be expressed, analogous to relation (4.19), as

$$n_c L_c \Delta P_o \; \alpha \; W^{3/4} \qquad (5.19)$$

where n_c and L_c are now considered to denote the number and length of characteristic capillaries of the pulmonary system, and where ΔP_o denotes the difference between the oxygen partial pressure immediately outside and inside the capillary vessel. We saw in Chapter 4 that the oxygen partial pressure of the blood inside a capillary varies with animal weight raised to the negative one-twelfth power. The same may be assumed to be true immediately outside the capillary, as considered in Chapter 4 for the systemic circulation. The difference ΔP_o will then also vary in this manner, and the four relations of Chapter 3 mentioned above and this last relation then allow us to determine relations for the pulmonary vessels. Such considerations show, in fact, that these scaling relations are identical to those found for the systemic vessels in Chapter 3. Thus, we need make no distinction between the scaling relations of the systemic and pulmonary vessels.

We have previously seen evidence supporting the validity of the scaling relations for the characteristic capillaries of the systemic system. We now consider additional evidence based on reported measurements of the net lateral surace area S_c and volume of B_c of the pulmonary capillaries. According to the present theory, these quantities are described in terms of the number n_c, radius r_c, and length L_c of characteristic capillaries by the equations

$$B_c = N_c(\pi r_c^2 L_c), \; S_c = n_c(2\pi r_c L_c) \qquad (5.20)$$

With these relations, the characteristic capillary radius r_c and net length $n_c L_c$ are then expressible as

$$r_c = \frac{2B_c}{S_c}, \; n_c L_c = \frac{S_c^2}{4\pi B_c} \qquad (5.21)$$

Measurements of the volume B_c and area S_c have been reported by Gehr et al. (1981) for a large number of mammals. Choosing typical mammals from this study and calculating the capillary radius using the above formula, we find the values shown in Fig. 5.11. It can be seen that these results clearly demonstrate the trend of increasing radius with increasing animal size, in qualitative agreement with that expected from the present theory. According to this theory, the radius should, in fact, vary with animal weight raised to the one-twelfth (0.083) power. Assuming this variation and using the data to calculate an average coefficient in the equation, we find

$$r_c = 0.0027 \; W^{1/12} \qquad (5.22)$$

This relation is illustrated graphically in Fig. 5.11 where it can be seen that it does, indeed, provide a remarkably good representation of the

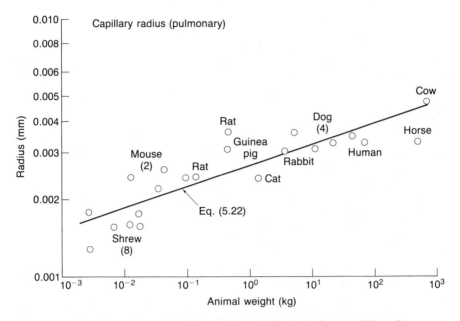

Fig. 5.11 Comparison of capillary radius with theoretical one-twelfth scaling law in the form of Eq. (5.22). Data derived from measurements of Gehr et al. (1981).

measurements. Interestingly, a best-fit power-law relation of the data provides a coefficient of 0.0027, identical to that of Eq. (5.22), and an exponent of 0.079, in close agreement with the one-twelfth value.

In a similar way, we may use the second of Eq. (5.21) to determine the net capillary length for various mammals. These results are shown in Fig. 5.12, where it is seen that there is an appreciable dependence on animal size. This is in general agreement with that expected from the present theory, since the relevant relations of Chapter 3 give the scaling relation for the net capillary length in the form

$$n_c L_c \ \alpha \ W^{5/8} W^{5/24} \ \alpha \ W^{5/6}$$

that is, in the form of animal weight raised to the five-sixths (0.83) power. Assuming this relation and using the data to calculate an average coefficient, we then find the equation

$$n_c L_c \ = \ (1.92 \ \times \ 10^5) \ W^{5/6} \qquad (5.23)$$

This equation is illustrated graphically in Fig. 5.12 where, like that for the capillary radius, it can be seen to provide a remarkably good representation of the measurements. A best-fit power-law relation provides, in this case, a coefficient of 1.76 x 10^5 and an exponent of 0.85, both in good agreement with the values of Eq. (5.23).

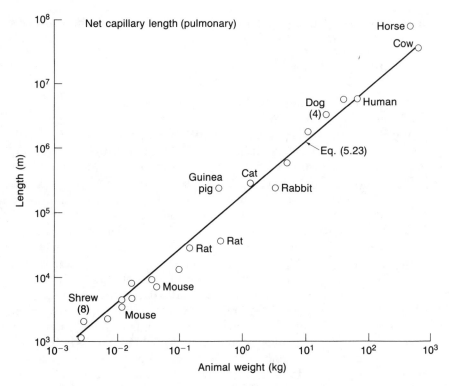

Fig. 5.12 Comparison of net capillary length with the theoretical five-sixths scaling law in the form of Eq. (5.23). Data derived from measurements of Gehr et al. (1981).

The above results for capillary radius and net length are therefore seen to provide further support for the validity of the scaling relations for the characteristic capillaries as predicted by the present theory. We may note also that Gehr et al. (1981) provided a best-fit power-law relation for their measurements of net capillary volume which showed this volume to vary directly with animal weight. This is in agreement with the assumptions of the present theory. From the scaling relations of Chapter 3, we have the result, directly, that

$$B_c \; \alpha \; n_c r_c^2 L_c \; \alpha \; W^{5/8} W^{2/12} W^{5/24} \; \alpha \; W \tag{5.24}$$

that is, that net capillary volume is directly proportional to animal weight. Such a condition was also established explicitly by the second of relations (3.20).

Similarly, the measurements of the net lateral area of the capillaries was shown by best-fit methods to vary with animal weight raised to the

0.95 power, a result also in good agreement with the present theory. In particular, the theoretical scaling relation is

$$s_c \ \alpha \ n_c r_c L_c \ \alpha \ W^{5/8} W^{1/12} W^{5/24} \ \alpha \ W^{11/12} \tag{5.25}$$

so that the lateral area is predicted to vary with animal weight raised to the eleven-twelfths (0.92) power.

We may note in passing that the variation of capillary radius with animal size, predicted by the present theory and indicated in Fig. 5.11, might be considered to imply a similar variation in size of the red blood cells, described in Chapter 2. This is particularly tempting since the radius of these disc-like cells for any mammal is of the same order of magnitude as the capillary radius. Such a variation is, however, not generally found to be true with existing measurements (Schmidt-Nielsen 1984), although there are special cases supporting the variation. For example, consistent with the trend shown in Fig. 5.11, the red blood cell of the dog has a radius of about 0.0035 mm, that of the human has a radius of about 0.0037 mm, and that of the elephant has a radius of about 0.0046 mm. In contrast, the red blood cell of the shrew has a relatively large radius of 0.0033 mm, while that of sheep have the relatively small radius of 0.0024 mm, and that of the horse has the relatively small radius of about 0.0028 mm. Thus, no general trend is indicated by these and similar measurements. The explanation would appear to be that factors other than capillary radius govern the specific sizing of the red blood cells and that capillary radius only sets a rough upper bound. Presumably, the high flexibility of the cells allows them to deform and pass through capillaries even when these sizing factors require relatively large cells in comparison with the capillary opening.

Oxygen Transfer

We now turn our attention to some other aspects of lung design that are associated with its role in oxygen transfer. Based on the scaling relation for the oxygen-consumption rate of mammals, we expect that the rate of oxygen absorption in the lungs will vary with animal weight raised to the three-fourths power and that the amount absorbed per breath will vary with the ratio of this rate to the breathing rate, that is, directly with animal weight. As a result, we may expect that the volume of air inspired and expired with each normal breath (the tidal volume) will also vary directly with animal weight. Finally, if lung volume is proportional to tidal volume for mammals of different size, we see that total lung volume should likewise vary directly with animal weight. Such a condition has, in fact, been found to be the case by Tenney and Remmers (1963) and more recently by Gehr et al. (1981). Some typical measurements are shown in

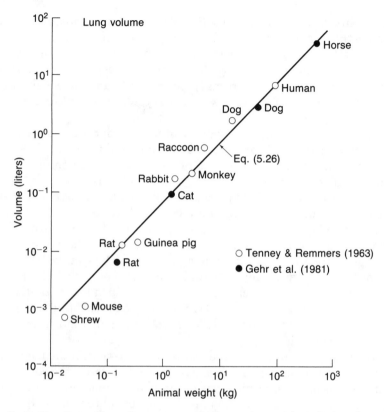

Fig. 5.13.   Comparison of measurements of lung volume with predictions from Eq. (5.26).

Fig. 5.13. The solid line passing through the data-points indicates predictions based on an assumed proportional relation between lung volume and animal weight. The equation, with its coefficient determined from the data, is expressible as

$$B_L = 0.068W \tag{5.26}$$

where B_L denotes the lung volume as expressed in liters.

The volume referred to above is equal, approximately, to the total lung capacity, that is the maximum volume of the lungs when fully expanded. With each breath, the resulting tidal volume of air represents only about 10% of the total volume of air in the lungs. Also, with each breath, there is a certain volume of air that fills the respiratory passages and is unavailable for gas exchange with the pulmonary capillaries. This is referred to as the dead-space volume and amounts to about 30% of the tidal volume. The alveolar air available for gas exchange with the blood is thus the difference between the tidal volume and the dead-space volume. This, like the tidal volume itself, is directly proportional to animal weight for mammals of varying size. The alveolar ventilation rate is then the product of the above quantity of alveolar air and the respiratory rate. The alveolar ventilation rate accordingly scales with animal weight to the three-fourths power, consistent with the oxygen-consumption rate of the body.

Alveolar ventilation is obviously of major importance in determining the concentration of oxygen and carbon dioxide in the alveoli and, hence, in controlling the exchange of these gases with the pulmonary blood by diffusion processes. Driving the diffusion is, of course, the partial pressures of the gases. Typical atmospheric air has an oxygen partial pressure of about 160 mm Hg and a carbon dioxide partial pressure of about 0.3 mm Hg. In contrast, in the alveolar air of humans, the oxygen partial pressure is decreased to about 104 mm Hg and the carbon dioxide partial pressure is increased to about 40 mm Hg (Guyton, 1971). The reasons for the differences are: (1) that alveolar air is only partially replaced by atmospheric air with each breath; (2) that oxygen is constantly being absorbed from the alveolar air and carbon dioxide is constantly being discharged into it; and (3) than the alveolar air has a greater moisture content than atmospheric air.

In considering details of the diffusion of gas across the membranes of the numerous alveolar sacs, it is customary to refer to the collection of membranes as a single respiratory, or pulmonary, membrane. From a design perspective, the area of this respiratory membrane must be chosen consistent with the rate at which oxygen and carbon dioxide must pass through it. However, since the diffusion of carbon dioxide from pulmonary blood to alveolar air is known to occur at a much faster rate than that of oxygen, the limiting design condition must involve consideration of the rate of oxygen transfer across the membrane.

The diffusion of oxygen through the respiratory membrane must, of course, be followed by diffusion through the net surface area of the pulmonary capillaries, and this must, in turn, be equal to the oxygen-consumption rate Q_o of the body. With these observations, we may write the following diffusion relations representing the oxygen transfer:

$$Q_o \; \alpha \; S_c \frac{\Delta P_o}{h_c} \; \alpha \; S_m \frac{\Delta P_{om}}{h_m} \tag{5.27}$$

where S_c and S_m denote the net diffusing surfaces of the capillaries and respiratory membrane, respectively. Here also, ΔP_o denotes the average difference in oxygen partial pressure on opposite sides of the capillary wall, of thickness h_c, and ΔP_{om} denotes the average difference in oxygen partial pressure, on opposite sides of the membrane, of thickness h_m. If we now assume the average partial pressure increases uniformly from the capillary blood to the alveolar air in contact with the respiratory membrane, we then have the additional relation

$$\frac{\Delta P_o}{h_c} = \frac{\Delta P_{om}}{h_m} \qquad (5.28)$$

Using this equation, we may then conclude from relation (5.27) that

$$S_c \; \alpha \; S_m \qquad (5.29)$$

that is, that the net capillary area and the respiratory-membrane area are proportional under change in animal size. Such a condition has, in fact, been found by Gehr, et al. (1981). This result accordingly supports the uniform pressure assumption and the associated pressure condition described by Eq. (5.28). It also gives us the design condition on the membrane area requiring that it should vary like net capillary area, that is, as animal weight raised to the eleven-twelfths (0.92) power of relation (5.25).

Measurements of respiratory membrane area by Gehr et al. (1981) have provided an empirical exponent of animal weight of 0.95, in close agreement with the present theoretical result of 0.92. Interestingly, earlier work by Tenney and Remmers (1963) had suggested that respiratory membrane area was proportional to oxygen-consumption rate and, hence, essentially porportional to animal weight raised to the three-fourths power. The difference has been traced by Gehr et al. (1981) to the inclusion of relatively large marine animals in the Tenney-Remmers study. These mammals have specialized respiratory apparatus which provided bias to the overall set of measurements. When excluded, the remaining terrestrial mammals of the study give scaling results consistent with those described here.

If we again consider relation (5.27), and use the theoretical scaling relations for the oxygen consumption rate and net capillary area, we see that the change in oxygen partial pressure per unit of capillary wall thickness must scale with animal weight to the negative one-sixth power. Because of the above assumption of uniform increase in partial pressure with distance from capillary blood, this same scaling relation must apply to the difference between the oxygen partial pressure of the air in contact with the respiratory membrane; and the oxygen partial pressure of the blood, when divided by the distance between the two, must scale in this same way.

Now, we ask, how does this distance scale? According to the present theory, the wall thickness of the capillaries must scale like their radius, that is, with animal weight raised to the one-twelfth power. Because of the similarity in the surface areas of the capillaries and respiratory membrane, we may reasonably assume the same scaling for the thickness of both. This will ensure that the weight of the respiratory membrane will vary directly with animal weight, consistent with the scaling of the lung weight indicated earlier in Table 5.2. In this case, neglecting any interstitial space, or assuming it to scale similarly, we then see that the total diffusion distance between the blood and alveolar air must scale as animal weight to the one-twelfth power. Interestingly, Gehr et al. (1981) found from measurements an exponent of 0.05, in reasonable agreement with the one-twelfth (0.08) value.

From the above two results, it now follows that the difference in the corresponding oxygen partial pressures must scale as animal weight to the negative one-twelfth power. The average oxygen partial pressure in the blood is already known to obey this relation, so this condition requires that it apply also to the oxygen partial pressure of the alveolar air in contact with the respiratory membrane. This is an interesting condition, because it suggests that the oxygen partial pressure in direct contact with the respiratory membrane will not generally have the accepted inflow value of about 104 mm Hg. If, for example, this condition is met for, say, a small animal such as a mouse or shrew of 0.010-kg weight, it would not be met for larger animals. For the human of 70-kg weight, it would be only about 50 mm Hg. Since the oxygen partial pressure of the alveolar air away from the membrane surface is higher, presumably that of the inflow value, the implication is that a non-uniform distribution of oxygen must exist in the alveolar sacs of, at least, all but the smallest mammals as a result of the diffusion process. This non-uniform distribution, referred to as stratification, has been discussed previously by Weibel et al. (1981).

5.5 MECHANICAL DESIGN OF THE HEART

We have previously considered various aspects of the mechanical design of the heart. In Chapter 1, we observed that the weight of the heart varies directly with animal weight. In Chapter 2, we saw that the mechanics of cardiac muscle contraction could be likened to that of an elastic spring with changing initial length. We have also seen in the theoretical considerations of Chapter 3 that the linear dimensions of the ventricles, and hence the heart as a whole, must vary with animal weight raised to the one-third power. In addition to the geometric scaling, we know from theoretical considerations, as well as measurements, that the resting heart rate decreases with increasing size, as animal weight raised to the negative one-fourth power. We now consider some additional relations associated with the design of the heart.

Ventricular Stress

To investigate the scaling relation for ventricular stress, that is, the force per unit of area acting on the ventricular walls, we write the ventricle-force relation, given by Eq. (3.1), in the form

$$\frac{F'}{h} = P \left(\frac{a}{h} - \frac{U}{h} \right) \tag{5.30}$$

where F' denotes the force per unit length of ventricular wall, a the radius of the ventricle, U its displacement at any time in the pumping cycle, and P the net internal pressure. The ratio F'/h has units of force per area and is seen to represent the stress in the ventricular walls. Now, because of the similarity in the pumping cycle, the displacement U will be proportional under size change to the amplitude value U_o. Recalling from arguments given in Chapter 4 that ventricular pressure does not change with animal size and, from results given in Chapter 3, that the ratios U_o/h and a/h do not change with size, we thus see that the ventricular stress must likewise remain unchanged under size change.

This result is significant because it means that the maximum stress developed in the ventricular walls during pumping is the same for the mouse as, say, for the elephant. Since rupture of the ventricular tissue is determined by a critical stress level, each animal can therefore operate at some fixed safe percentage of the rupture strength. In engineering terms, we say that the heart of the mouse and elephant are both designed to operate during rest with the same reserve and with the same factor of safety against stress failure.

The condition just described is simple in concept, yet far-reaching in its implications regarding the design of the cardiovascular system. It may, in fact, be a principal reason that similar systems exist among mammals of vastly different size and that scaling relations can be specified for the system and system performance. The requirement of size-independent ventricular stress can only apply if the blood pressures of mammals are independent of size. Moreover, this can only be the case if the viscous resistance in the small blood vessels and the inertial resistance in the large vessels are independent of animal size. These requirements, together with considerations of heart excitation and diffusion as described in Chapter 3, thus govern the sizing of the vessels and the heart rate among mammals of different size. As we have seen, these are fundamental in prescribing the various characteristics of the system for any animal size.

Work and Efficiency

The work W_k of a heart ventricle may be defined as the product of the ventricle-wall force F' and the inward displacement U. From Eq. (5.30), we may express this work as

$$W_k = F'\ell\, U = \ell h P U \left(\frac{a}{h} - \frac{U}{h}\right) \tag{5.31}$$

where ℓ denotes the ventricular length.

By the same general arguments as above, we can see from this equation that the ventricular work must scale with the product $\ell h U$. From the relevant relations of Chapter 3, this is further seen to scale directly with animal weight. Thus, the work of a ventricle, and, hence, that of the heart also, must vary directly with animal weight.

The power output P_w of the heart is the rate at which it performs work. This may therefore be considered to be proportional to the product of the work and heart rate, that is,

$$P_W \ \alpha \ \omega W_k \tag{5.32}$$

From the known scaling law for the heart rate and the above scaling relation for the work, we see that the power must scale with animal weight raised to the three-fourths power. As discussed at the beginning of this chapter, this is also the manner in which the oxygen-consumption rate, or fuel-consumption rate, of the heart scales. The ratio of power output to oxygen consumption rate is thus seen to be independent of animal size.

This last ratio represents the mechanical efficiency of the heart and can be estimated for all mammals using known values for the human heart. Thus, as tabulated by Sagawa (1973), the normal power output for the human heart is about 12.5 kg-m/min., and the oxygen consumption rate is about 22 ml/min. The latter is equivalent to 45.3 kg-m/min. so that the normal effeciency of the heart is 12.5/45.3 or about 28%. During strenuous exercise, the power output of the heart increases more than the rate of oxygen consumption, and the efficiency increases to about 40%. These efficiencies are comparable to those of a small mechanical pump. The fact that efficiencies less than 100% exist in both of these cases is, of course, accounted for by the heat generated in the process.

REFERENCES

ADOLPH, E. F. 1943. *Physiological Regulations*. Lancaster, PA: The Jaques Cattell Press.

————.1949. Quantitative relations in the physiological constitution of mammals. *Science* 109: 579–85.

BRODY, S. 1945. *Bioenergetics and Growth, With Special Reference to the Efficiency Complex in Domestic Animals.* New York: Reinhold Publishing.

DAVSON, H., and M. G. EGGLETON, eds. 1968. *Principles of Human Physiology.* Philadelphia: Lea and Febiger.

GEHR, P., D. K. MWANGI, A. AMMANN, G. M. D. MALOIG, C. R. TAYLOR, and E. R. WEIBEL. 1981. Design of the mammalian respiratory system. V. Scaling morphometric pulmonary diffusing capacity to body mass: wild and domestic mammals. *Resp. Physiol.* 44: 61–86.

GUYTON, A. C. 1971. *Textbook of Medical Physiology.* Philadelphia: W. B. Saunders.

KUNKEL, P. A., JR. 1930. The number and size of the glomeruli in the kidney of several mammals. *Bulletin of the Johns Hopkins Hospital* 47(5): 285–91.

MARTIN, A. W., and F. A. FUHRMAN. 1955. The relationship between summated tissue respiration and metabolic rate in the mouse and dog. *Physiol. Zool.* 28: 18–34.

MUNRO, H. N., and E. D. DOWNIE. 1964. Relationship of liver composition to intensity of protein metabolism in different mammals. *Nature* 203: 603–04.

PAPPENHEIMER, J. R., E. M. RENKIN, and L. M. BORRERO. 1951. Filtration, diffusion and molecular sieving through peripheral capillary membranes. *Am. J. Physiol.* 167(1): 13–46.

PROTHERO, J. W. 1982. Organ scaling in mammals: the liver. *Comp. Biochem. Physiol.* 71A(4): 567–77.

SAGAWA, K. 1973. The heart as a pump. In *Engineering Principles in Physiology* Vol. II, J. H. Brown and D. S. Gain, eds., pp. 101–26. New York: Academic Press.

SCHMIDT-NIELSEN, K. 1984. *Scaling: Why is Animal Size so Important?* Cambridge: Cambridge University Press.

SMITH, R. E. 1955. Quantitative relations between liver mitochondria metabolism and total body weight in mammals. *Ann. N. Y. Acad. Sci.* 62: 405–21.

STAHL, W. R. 1967. Scaling of respiratory variables in mammals. *J. Appl. Physiol.* 22(3): 453–60.

TENNEY, S. M., and J. E. REMMERS 1963. Comparative quantitative morphology of the mammalian lung: diffusing area. *Nature* 197: 54–56.

WEIBEL, E. R., C. R. TAYLOR, P. GEHR, H. HOPPELER, O. MATHIEU, and G. M. O. MALOIY. 1981. Design of the mammalian respiratory system. IX. Functional and structural limits for oxygen flow. *Resp. Physiol.* 44: 151–64.

6

Cardiac Response
Characteristics

Up to this point, we have used the design theory of Chapter 3 to establish scaling laws governing changes in cardiovascular variables with change in mammal size. The success of this work, as seen in the previous two chapters, indicates the general applicability of the theory for this purpose. We now examine the further applicability of the theory in describing certain detailed response characteristics of the system.

6.1 SIMPLIFIED GOVERNING EQUATION

The governing design equation in the present theory is Eq. (3.12), which relates the inward displacement (U), velocity $(\dot{U})$, and acceleration $(\ddot{U})$ of the heart ventricle to the periodic contractile force in its walls. Our present goal is to use this equation to determine the theoretical variation of ventricular volume, outflow, and pressure during a cycle of pumping. For this purpose we will find it convenient to simplify the equation by neglecting all terms containing products of U, $\dot{U}$, and $\ddot{U}$ such as $U\dot{U}$, $U\ddot{U}$, etc. Such a simplification is strictly valid only when the ventricular displacements are very small in comparison with the mean ventricular radius, but can be regarded as approximately correct for the actual case where they are typically less than 15% of the mean radius. We also assume here that the force variation in the ventricular walls is described by a simple sinusoidal function.

With the above restrictions, we may write the governing equation in simplified form as

$$M\ddot{U} + C\dot{U} + KU = F_o \sin \omega t \qquad (6.1)$$

where M, C, and K denote effective inertia, friction, and elastic coefficients of the system, F_o denotes the amplitude of the contractile force, ω denotes the contractile frequency, equal to the heart rate in radians per second, and t denotes time. Note that, with ω expressed in radians per second, the ratio $\omega/2\pi$ equals the heart rate in beats per second, and the ratio $60\omega/2\pi$ equals the heart rate in beats per minute.

It is instructive to interpret the products on the left-hand side of Eq. (6.1) in terms of fractions of the total contractile force (per unit of ventricle length) described by the product on the right-hand side. Thus, the first product, $M\ddot{U}$, represents that fraction, or part, of the ventricular force needed to overcome the inertial resistance of the blood in accelerating it through the main connecting vessels, the second term, $C\dot{U}$, represents that part of the ventricular force needed to overcome the viscous resistance of the blood in moving it through the characteristic capillary vessels, and the third term, KU, represents that part of the ventricular force needed to overcome the elastic resistance to contraction of the ventricle itself. As the equation indicates, the sum of these three parts must equal the total amount of contractile force available. The relative magnitude of each depends, of course, on the instantaneous values of the inward acceleration, velocity, and displacement of the ventricular walls.

Eq. (6.1) provides us with an idealized description of the contractile part of the cardiac cycle, but gives us no information on the expansion part when the output valve to the vascular system is closed. During this interval, blood drains into the ventricle from the adjacent atrium and the ventricle expands under its own restoring forces. However, as noted in Chapter 3, near the end of this expansion, the atrium contracts, squeezing additional blood into the ventricle and completing its expansion. Because of this, the starting conditions for the next cycle of pumping are identical to those of the previous one and are independent of the details of the expansion.

Consistent with Eq. (6.1) and this last condition, we may express the time-dependent relations for the contractile displacement U, velocity $\dot{U}$, and acceleration $\ddot{U}$ in the well-known periodic forms:

$$U = U_o \sin (\omega t - \phi)$$

$$\dot{U} = U_o \, \omega \, \cos (\omega t - \phi) \qquad (6.2)$$

$$\ddot{U} = -U_o\omega^2 \sin (\omega t - \phi)$$

where expressions for the displacement amplitude U_o and phase angle ϕ are φ determined from Eq. (6.1) as

$$U_o = \frac{F_o/K}{\left\{ \left[1 - \left(\frac{\omega}{\Omega} \right)^2 \right]^2 + \left[\frac{\zeta}{\Omega} \frac{\omega}{\Omega} \right]^2 \right\}^{1/2}} \tag{6.3}$$

$$\tan \phi = \frac{\dfrac{\zeta}{\Omega} \dfrac{\omega}{\Omega}}{1 - \left(\dfrac{\omega}{\Omega} \right)^2} \tag{6.4}$$

with Ω and ζ defined by

$$\Omega = \sqrt{\frac{K}{M}}, \ \ \zeta = \frac{C}{M} \tag{6.5}$$

Equation (6.2) describes the variation of the inward ventricular motion with time during the pumping interval. The associated descriptions of the time variation of ventricular volume B_v about its mean value B_o and the instantaneous outflow of blood Q are given by

$$B_v = B_o - \bar{S} U_o \sin (\omega t - \phi) \tag{6.6}$$

and

$$Q = \bar{S} U_o \omega \cos (\omega t - \phi) \tag{6.7}$$

where $\bar{S} = 2\pi a \ell$ denotes the mean inside surface area of the ventricle, a being its mean inside radius and ℓ its length.

From Eq. (3.10), we may also write the expression for the time variation of ventricular pressure during pumping as

$$P = P_s + P_{10} \cos(\omega t - \phi) - P_{20} \sin (\omega t - \phi) \tag{6.8}$$

where the pressure coefficients P_{10} and P_{20} are given by

$$P_{10} = \frac{C U_o \omega}{a}, \ P_{20} = \frac{M U_o \omega^2}{a} \tag{6.9}$$

6.2 PREDICTIONS FOR THE RESTING HUMAN

It is of interest to examine the above theoretical solutions for ventricular response using typical values for the left ventricle of the human in the resting state. From the ventricle-volume curve given in Fig. 2.9, we may see that approximate values for mean volume B_o and stroke amplitude $\bar{S} U_o$ are $B_o = 100$ ml and $\bar{S} U_o = 35$ ml. From the ventricular pressure curve of this same figure, we may next see that the pressure at the beginning of contraction is about 100 mm Hg, and at the end it is about 85 mm Hg. Using Eq. (6.8), it can therefore be determined that these values correspond approximately to values of P_s = 93 mm Hg and P_{20} = 8 mm Hg. Consistent with the maximum pressure of 120 mm Hg, we also choose P_{10} = 26 mm Hg. Finally, we take ω = 7.33 radians/sec, corresponding to a heart rate of 70 beats/min.

With these results, we then find the ventricle-response characteristics during contraction described by the equations

$$B_v = 100 - 35 \ \sin(\omega t - \phi)$$

$$Q = 257 \ \cos(\omega t - \phi) \tag{6.10}$$

$$P = 93 + 26 \ \cos(\omega t - \phi) - 8 \ \sin(\omega t - \phi)$$

These relations are illustrated in Fig. 6.1 for two full pumping cycles. It can be seen that the general characteristics of the response are similar to the actual ones shown earlier in Fig. 2.9. During ventricular contraction, the outflow of blood rises to a maximum value and then falls to zero at the end of the contraction period. The ventricular pressure also rises during contraction to a maximum value of about 120 mm Hg and then falls to the value of 85 mm Hg during the last stages of contraction. Interestingly, the difference between the pressures at the beginning and end of contraction is, according to the present theory, due solely to the inertial resistance of the blood.

The above equations give no information on the response during ventricular expansion. However, the ventricular volume at the beginning of each cycle of contraction must be the same, so that the expansion must vary with time in the general manner indicated by the dashed line in Fig. 6.1. The outflow must also be zero during expansion, and the ventricular pressure must continue to fall to near zero during expansion, both as indicated.

For comparison purposes, the pressure variation in the aorta has been indicated by dotted lines in Fig. 6.1. This variation has been estimated from that shown earlier in Fig. 2.9, and illustrates the familiar 120/80 range of pressures over a heart cycle. The small rise in pressure immediately following the end of ventricular contraction (and value closure) is due to radial contraction of the aorta from its prior expansion under the action of ventricular pressure.

It may be noticed that the peak outflow value indicated in Fig. 2.9 is about twice that given by the present description. There are two main reasons for this discrepancy. One is that the time interval of outflow is smaller in the actual case than in the theoretical one. The other is that the actual output curve is more triangular shaped than the theoretical one. For both of these conditions, the actual peak outflow must, of course, be

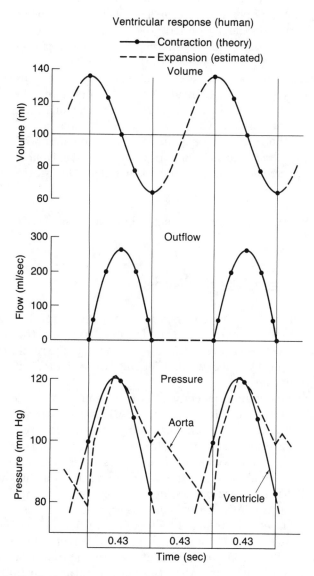

Fig. 6.1   Predictions of ventricular volume, outflow, and pressure from simplified theory

For both of these conditions, the actual peak outflow must, of course, be greater than the theoretical one if both are based on the same amount of blood being pumped per heart cycle. This difference of peak values indicates a limitation on the simplified governing equation considered here, but not necessarily on the more general governing equation established in Chapter 3.

In spite of the above difference in peak outflow, the present theory does give an excellent representation of the average blood flow. This is determined by the ratio of the total area under the outflow curve during contraction to the period of the cardiac cycle. The area is easily estimated from either of the outflow curves in Fig. 6.1 to be about 71 ml, and the period of the cardiac cycle is seen to be about 0.86 sec. Hence, the average blood flow, or cardiac output, given by the present description is

$$Q_b = 71/0.86 = 83 \text{ ml/sec} = 5000 \text{ ml/min.}$$

which agrees with the generally accepted values (5000 – 5500 ml/min) for man.

6.3 CARDIAC RESPONSE TO EXERCISE

The above considerations have shown us that the simplified description given in this chapter can provide a good representation of the cardiac output and ventricle-pressure variation of the human in the resting state. We now examine what the theory can offer regarding changes in ventricular response during exercise.

From ordinary experience, we know, of course, that heart rate increases during exercise. Measurements during very strenuous exercise show, in fact, that heart rate can increase by a factor of two or so, and that cardiac output can increase by a factor of about five. Some original measurements of this kind (Bock et al., 1928, as described in Davson and Eggleton 1968) are shown in Fig. 6.2, where cardiac output, heart rate, and venous saturation are plotted against oxygen-consumption rate for increasing amounts of exercise on a bicycle ergometer. At rest, the cardiac output is seen to equal the usual value of about 5 l/min. while, at maximum effort, it is about 25 l/min. The heart rate at rest is likewise seen to equal the typical value of about 70 beats/min. while that at maximum effort is about 160 beats/min. Also, the venous saturation falls from its value of about 75% at rest to about 45% during maximum effort.

We may recall from our earlier discussion in Chapter 1 that cardiac output can be expressed as the product of heart rate, in units of beats per minute or beats per second, and blood volume pumped per cycle, or stroke volume, in units of liters or milliliters. From these results, we see that the heart-rate increase during strenuous exercise can account for only a part of the increase in cardiac output and that increased stroke volume must account for the rest. Our attention is therefore drawn initially to what the present theory can tell us regarding the details of the increased cardiac output accompanying increased heart rate.

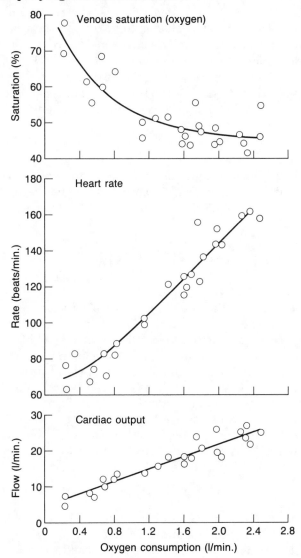

Fig. 6.2 Measurements of venous saturation, heart rate, and cardiac output of man during exercise with varying rates of oxygen consumption. Data source: Davson and Eggleton (1968).

Cardiac Output

An inspection of Eq. (6.6) will show us that the maximum ventricular volume over a cardiac cycle will equal the sum $B_0 + \bar{S}U_o$ and the minimum volume will equal $B_0 - \bar{S}U_o$, where B_0 denotes the mean ventricular volume, $\bar{S}$ denotes the mean inside ventricular surface area, and U_0 denotes ventricular displacement. The change in volume of the ventricle over a pumping cycle (the stroke volume) is thus equal to $2\bar{S}U_o$. The cardiac output is then determined by the product of this volume and the heart rate in beats per unit of time. Since ω denotes heart rate in radians per second in the formulas of this present chapter, the rate in beats per second is $\omega/2\pi$. Thus, cardiac output (in volume per second) is expressible as

$$Q_b = \frac{1}{\pi}\,\bar{S}U_o\omega \tag{6.11}$$

Combining this last equation with the expression for the displacement amplitude U_o given by Eq. (6.3), we find that cardiac output is also expressible as

$$\frac{Q_b}{Q_{bo}} = \frac{\omega/\Omega}{\left\{\left[1 - \left(\frac{\omega}{\Omega}\right)^2\right]^2 + \left[\frac{\zeta}{\Omega}\frac{\omega}{\Omega}\right]^2\right\}^{1/2}} \tag{6.12}$$

where $Q_{bo} = \bar{S}\Omega F_o/\pi K$.

Numerical examination of Eq. (6.12) will show that the predicted cardiac output will reach a maximum value when the heart-rate ratio ω/Ω equals unity, corresponding to the so-called resonant condition in mechanical vibrations. Taking the previously noted maximum increase in heart rate during strenuous exercise to be associated with this resonant condition, we have $\omega = \Omega = 16.8$ radians/sec (160 beats/min.). The resting state, with heart rate = 7.33 radians/sec (70 beats/min.), must then correspond to a heart-rate ratio ω/Ω of 7.33/16.8 or about 0.44. From the definitions of the friction factor ζ, given by Eq. (6.5), and the pressure coefficients in Eq. (6.8), we also find the friction-ratio ζ/Ω expressible as

$$\frac{\zeta}{\Omega} = \frac{P_{10}}{P_{20}}\frac{\omega}{\Omega} \tag{6.13}$$

Using the value $P_{10}/P_{20} = 3.3$, as determined from the pressure coefficients of the previous section, and the value $\omega/\Omega = 0.44$, we then determine the friction ratio as $\zeta/\Omega = 1.5$. With these results and the assumption of a typical value of resting cardiac output of 5 l/min., we may next evaluate the constant Q_{bo} in Eq. (6.12) as $Q_{bo} = 11.9$ l/min.

We are now in a position to use Eq. (6.12) to predict the theoretical variation of cardiac output with frequency of heartbeat. There are two cases that are worthwhile to consider, that where the friction factor ζ/Ω is held fixed at the resting value, and that where it is decreased to account for the known opening of additional capillary vessels during exercise. The first case would describe, for example, the variation in cardiac output that results when increased heart rate is caused by artifical pacers, while the second would describe the variation that results when the increased heart rate is due to exercise. In the latter case, we assume for the strenuous-exercise data in Fig. 6.2 that the flow resistance is reduced by a factor of about three. This corresponds, in the present theory, to a tripling of the active capillaries over those working in the resting state and would require a friction factor $\zeta/\Omega = 0.5$, that is, one-third the resting value.

The variations of cardiac output with heart rate for these two cases have been calculated using Eq. (6.12) with $\Omega = 16.8$ radians/sec (160

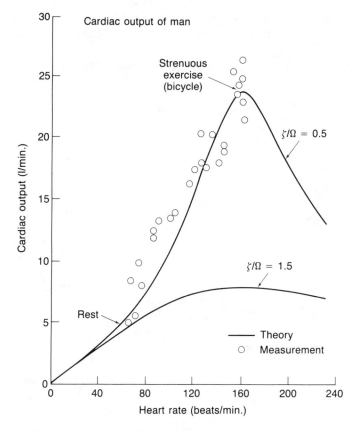

Fig. **6.3** Comparison of theoretical prediction of cardiac output with measurements of Fig. 6.2. The upper curve is for the case of exercise where increase in heart rate is accompanied by opening of reserve capillaries. The lower curve is that corresponding to increased heart rate only.

beats/min.). The results, with heart rate converted to beats per minute, are shown in Fig. 6.3, together with the measurements given earlier in Fig. 6.2. It can be seen that the maximum value of the cardiac output in exercise is predicted to be about five times that of the resting state, consistent with the measurements. Interestingly, it can also be seen that the measurements tend to follow the predictions for the friction ratio $\zeta/\Omega = 0.5$ as the heart rate increases from its rest value to that at strenuous exercise. Although there is some scatter in the measurements, the trend can be regarded, within the context of the present theory, as indicating that all reserve capillaries are active for exercise states ranging from mild to strenuous.

Another interesting aspect that can been seen from the results presented in Fig. 6.3 is that increased heart rate without the accompanying increase in the number of active capillaries gives rise to only a relatively small increase in cardiac output from the rest state. This is consistent with studies where artifical pacers have been used to increase heart rate of resting patients (Sagawa, 1973).

Similar results are shown in Fig. 6.4 using measurements by Reeves

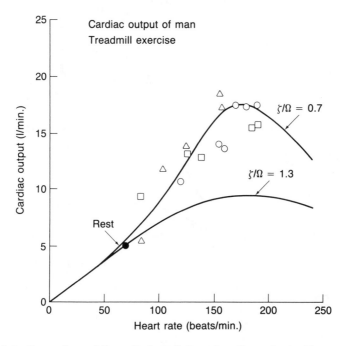

Fig. 6.4 Comparison of theoretical prediction of cardiac output with measurements for the case of treadmill exercise. The upper curve is for the case where increased heart rate is accompanied by opening of reserve capillaries. The lower curve is that for increased heart rate only. Data source: Reeves et al. (1961).

et al. (1961). for individuals engaged in treadmill walking at various grades. The most strenuous exercise in this case resulted in maximum heart rates of 160–190 beats/min. and oxygen-consumption rates of 2.3 – 2.4 l/min., consistent with the extreme values found in the data of Fig. 6.2 for bicycle exercise. However, the maximum cardiac output reported in the treadmill experiments is only about 18 l/min., in contrast with the value of about 25 l/min. observed in the measurements for bicycle exercise.

This difference between treadmill and bicycle exercise was noted also by Reeves et al. (1961) using a collection of data from various sources. Hence, it cannot be attributed to differences in individual test subjects. One explanation is that treadmill walking can lead to high heart rates and oxygen-consumption rates without causing the opening of reserve capillaries in all parts of the body where these exist. Thus, not all of these parts of the body may be affected in treadmill walking, in contrast with bicycle exercise, where they presumably are.

A comparison of the treadmill results with those predicted by Eq. (6.12) is shown in Fig. 6.4. The natural frequency Ω of the heart for these test subjects was estimated to be 18.5 radians/sec (175 beats/min.). The friction ratio for resting conditions was estimated in the same way as earlier to be 1.3, and the constant Q_{b_o} estimated to be 12.3 l/min. During exercise, the friction ratio was assumed, from consideration of the measured peak cardiac output, to be equal to 0.70. This corresponds to about a doubling of the number of active capillaries over those in the rest state in contrast with the factor of three found appropriate for bicycle exercise.

It can be seen that the agreement between the theoretical predictions and the measurements is good and again suggests, within the context of the present theory, that essentially all reserve capillaries in those parts of the body affected by the treadmill walking are open for the full range of exercise levels considered. Otherwise, a friction factor varying with exercise level would be needed to describe the measurements.

In summary, we have seen from the above work that the description provided by the present design theory indicates that cardiac output during strenuous exercise is the result of three factors: (1) increased heart rate, (2) increased ventricular displacement, and (3) decreased viscous resistance to blood flow. These three factors have, in fact, long been recognized as major contributors to increased cardiac output (Guyton, 1971). The explanation for the second factor provided by the present theory is, however, new and is indicated to be due to dynamic amplification, rather than to increased strength of cardiac contraction as is normally assumed. Thus, the present description requires no such increase. Instead, the amplification of the ventricular displacement is indicated to occur as a dynamic resonant phenomenon accompanying the increased heart rate. An examination of Eqs. (6.1) through (6.4) will show, moreover, that it arises because the elastic ventricular force and the inertial-resistance force of the blood tend to balance one another as the heart rate approaches the resonant value, leaving the cardiac contractile force to balance only the viscous resistance force of the blood. Hence, for the same strength of cardiac contraction, greater ventricular displacement, and outflow, can occur at and near the resonant heart rate than would otherwise be possible.

The present theory does not, of course, rule out the possibility of increased contractile strength from sources such as sympathetic nerve stimulation. Any increase would then need to be offset by a decrease in the number of capillary openings in order to describe the measurements in Figs. 6.3 and 6.4. If adequately understood, such a situation might, for instance, allow the assumption in the present theory of progressively increasing capillary openings with increasing exercise without loss of correlation with the cardiac-output measurements.

Ventricular Pressure

Let us next examine the ventricle-pressure characteristics during strenuous exercise. The pressure variation over a heart cycle is described in general terms by Eq. (6.8), with defining pressure coefficients P_{10} and P_{20} given by Eq. (6.9). For constant amplitude of contractile force F_o and constant inertia and elastic coefficients M and K appearing in Eqs. (6.3) and (6.5), we may express these coefficients as

$$P_{10} \ \alpha \ (\zeta/\Omega) \ (U_o/U_s)(\omega/\Omega) \tag{6.14}$$

$$P_{20} \ \alpha \ (U_o/U_s)(\omega/\Omega)^2 \tag{6.15}$$

where $U_s = F_o/K$. These results are important because they now allow us to scale the known pressure coefficients from the rest state to obtain values for the strenuous-exercise state.

Values for the right-hand side of these relations for the resting and exercise states are readily determined using Eq. (6.3) and previously established values for the necessary friction and heart rate parameters. For rest, we have, in particular, the ratios

$$\zeta/\Omega \ = 1.5, \ \omega/\Omega = 0.44, \ U_o/U_s = 0.96$$

and for strenuous exercise we have

$$\zeta/\Omega \ = 0.5, \ \omega/\Omega \ = 1, \ U_o/U_s = 2.0$$

Thus, we may use relations (6.14) and (6.15) to write

$$P_{10}(\text{exercise})/P_{10}(\text{rest}) = 1.58 \tag{6.16}$$

$$P_{20}(\text{exercise})/P_{20}(\text{rest}) = 10.8 \tag{6.17}$$

In our earlier work, we found P_{10}(rest) = 26 mm Hg and P_{20}(rest) = 8 mm Hg. Hence, using the above results, we have

$$P_{10}(\text{exercise}) = 41 \text{ mm Hg}$$

$$P_{20}(\text{exercise}) = 86 \text{ mm Hg}$$

With these coefficients, we then may express the ventricular pressure relations of Eq. (6.8) as

$$P = 93 + 41 \cos (\omega t - \phi) - 86 \sin (\omega t - \phi) \qquad (6.18)$$

where we have used the same static value of P_s = 93 mm Hg as found earlier for the resting state.

Predictions from this equation are plotted in Fig. 6.5 for one complete contractile interval and contrasted with those for the resting state (Eq. 6.10). The maximum pressure is seen to increase from 120 mm Hg to 188 mm Hg during exercise, consistent with general observations. In fact, during even the most severe exercise, systolic pressures are seldom found to rise much above 180 mm Hg.

It is also seen in Fig. 6.5 that the period of the contractile cycle in

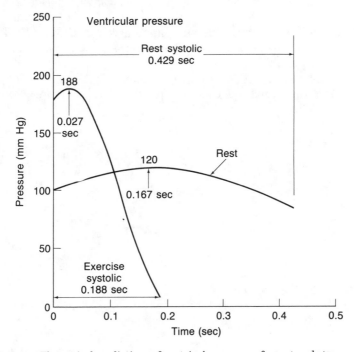

Fig. 6.5 Theoretical predictions of ventricular pressure for rest and strenuous exercise

exercise is less than half (0.44) that of the rest state as a direct result of the increased pulse rate. The ratio of the times of occurrence of the peak value in exercise and rest can, moreover, be seen to equal $0.026/0.167 = 0.16$. This indicates that the peak pressure develops much earlier in the cardiac cycle during exercise than during rest. A ratio of 0.44 would, of course, indicate similar times of occurrence. Additional calculations show, in fact, that the peak pressure in exercise occurs after only 10% of total ventricular contraction has taken place in contrast with the rest state, where about 35% of contraction occurs before the development of peak pressure. This effect is attributable to the increased importance of inertial resistance to blood flow on the ventricular pressure.

REFERENCES

BOCK, A. V., C. VANCAULAERT, D. B. DILL, A. FOLLING, and L. M. HURXTHAL. 1928. Studies in muscular activity. III. Dynamical changes occurring in man at work. *J. Physiol*. 66: 136–61.

DAVSON, H. and M. G. EGGLETON, eds. 1968. *Principles of Human Physiology*. Philadelphia: Lea and Febiger.

GUYTON, A. C. 1971. *Textbook of Medical Physiology*. Philadelphia: W. B. Saunders.

REEVES, J. T., R. F. GROVER, S. G. BLOUNT, JR., and G. FILLEY. 1961. Cardiac output response to standing and treadmill walking. *J. Appl. Physiol*. 16(6.2): 283–88.

SAGAWA, K. 1973. The heart as a pump. In *Engineering Principles in Physiology* Vol. II, J.H.U. Brown and D. S. Gain, eds., pp 101–26. New York: Academic Press.

7

Review and Reflection

It has been the purpose of this work to develop an engineering description of the design of the cardiovascular system of mammals which will allow prediction of well-known scaling laws for heart rate and oxygen-consumption rate of mammals. It has also been a goal to use this description, once firmly established, to investigate other aspects involved in the basic workings of the system. Having met these objectives in the previous chapters, we now review the main ideas presented and make some additional closing remarks on the overall results.

7.1 REVIEW AND SUMMARY

We began in Chapter 1 by considering measurements demonstrating the well-known conditions that the heart weight and blood weight of mammals varied directly with animal weight, while the heart rate and the oxygen-consumption rate varied with animal weight raised to the minus one-fourth and plus three-fourths powers, respectively. We then noted that the existence of these scaling relations, particularly the latter ones, implied that the design of the cardiovascular system of all mammals, ranging in size from, say, mice to elephants, was governed by the same basic design equations. It was next observed that previous efforts to establish these scaling relations from heat-balance arguments or modeling theory were inadequate, both in predicting the correct relations and in dealing with the conflicting scaling requirements of inertial and viscous resistance to blood flow and elastic-like heart response. The tasks required of the present work were thus seen to be the identification of a design description that would, indeed, place these scaling relations on a theoretical basis and the use of this description to investigate other interesting aspects of the system and system performance.

In Chapter 2, we reviewed the key engineering features of the system: the working fluid (the blood), the pumps (the heart), and the piping and diffusion apparatus (the blood vessels). Associated with this review were some discussions of the engineering aspects of the workings of its major components. Of particular significance was the description of cardiac muscle mechanics. This, it will be recalled, was based on a simple analogy with a non-linear elastic spring whose initial length is considered to shorten under stimulation. Also of significance was the explicit demonstration that viscous resistance to blood flow is dominant in the small blood vessels and that inertial resistance is dominant in the larger ones.

The foregoing results were used in Chapter 3 to construct an appropriate engineering description for the cardiovascular system which represented its essential parts, namely the pumping actions of the heart and the resulting blood flow through the vascular system.

A dimensional analysis of the variables in this description, together with consideration of the known proportional relations between heart weight, blood weight, and animal weight, led directly to the following scaling laws:

(a) All linear dimensions of the heart must scale with animal weight to the one-third power.

(b) The amplitudes of heart contractions and the force per unit length in its walls must scale with animal weight to the one-third power.

(c) Static or mean blood pressure in the heart is constant, independent of animal size.

(d) Heart rate and length of a main connecting blood vessel (aorta or pulmonary artery) are inversely proportional to one another under change in mammal size.

The analysis also provided three scaling relations involving the radius and length of the larger blood vessels (referred to as connecting vessels) and the radius, length, and number of the smaller vessels (referred to as characteristic capillary vessels). Two additional relations were thus needed to define the scaling relations for the geometry of these vessels. These were obtained from (1) considerations of the initiation and spread of the cardiac contraction, and (2) consideration of the diffusion of a substance into or out of a heart cell (or average body cell since heart weight and body weight are proportional under change of scale) during a cycle proportional to the cardiac cycle. With these results, the following scaling relations for the geometry of the vascular system were derived:

> The radius and length of connecting vessels in mammals vary with animal weight raised to powers of three-eighths and one-fourth, respectively, while the number remains unchanged. The radius, length, and number of characteristic capillary vessels vary with animal weight raised to powers of one-twelfth, five twenty-fourths, and five eighths, respectively.

This work therefore showed that scaling laws could indeed be derived theoretically for the cardiovascular system in the realistic situation where both viscous and inertial resistances to blood flow are taken into account, along with cardiac muscle elasticity. As indicated, the results require non-uniform scaling of the number and geometric dimensions of the two classes of vessels.

In addition to these scaling laws, it was found from the preceding considerations of the initiation and spread of cardiac contractions and the cell diffusion processes that the number of cardiac (or average body) cells must scale with animal weight raised to the five-eighths power and that the linear dimensions of these cells must scale with animal weight raised to the one-eighth power. Thus, cell geometry and vascular geometry were found to be related to one another. Of major importance also is the fact that these last relations led directly to the desired theoretical prediction that heart rate and oxygen-consumption rate must scale with animal weight raised to the minus one-fourth and plus three-fourths powers, respectively.

Having met the first objective of the work of devising an engineering description capable of predicting these known scaling laws, we next considered the application of the theory to other aspects of the system. In Chapter 4, we found that the above predicted scaling law for the linear dimensions of the heart was consistent with existing measurements, as was the above scaling law for the radius of connecting vessels. It was also shown that a three-fourths power scaling law was predicted for cardiac output, and that a size-independent rule applied to blood pressures, both of which were in agreement with general experience and existing measurements. Scaling laws were likewise developed from the theory for the oxygen partial pressure in mammalian blood and the capillary density in muscle tissue, and these, too, were shown to be in good agreement with existing measurements. Interestingly, the theory required these variables to vary with animal weight to the minus one-twelfth and minus one-sixth powers, respectively, so that smaller mammals are predicted to have proportionately higher oxygen partial pressures in their blood and proportionately high densities of capillaries in their muscles.

The theory was further applied in Chapter 5 to a study of the individual organs of the body. Attention was first directed to the oxygen-consumption rate of the organs, since the supply of oxygen is a major function of the cardiovascular system. Using published measurements on tissue respiration, this was shown to scale with animal weight raised to the three-fourths power for all organs, consistent with that for the body as a whole and consistent with organ blood flow as was discussed later. The concept of an average organ cell was introduced such that the product of the number and volume of these cells was proportional to organ weight. Using published data to determine approximate scaling relations for organ weights and combining these with the three-fourths scaling law for the oxygen-consumption rate, the scaling relations for the number and characteristic length of the average organ cells was established for several organs. The scaling relations for the rates of operation of the cells was also determined from diffusion considerations. Cell rates for the skin, lungs, and muscle were found to scale in the same way as heart rate, and the weights of these organs were found proportional to body weight under change of size. These cells were thus of the kind referred to in Chapter 3 as average body cells. In contrast, cell rates for the liver and kidneys were found to vary less than those for the average body cells under change in mammal size, and the rate of operation of brain cells was found to be constant, independent altogether of animal size.

We next considered average blood flow to the individual organs and showed that this should scale with animal weight in the same manner as cardiac output, consistent also with the scaling of the oxygen-consumption rate of the organs noted above. The accuracy of this prediction was shown using published data on blood flow to the liver.

Additional study in this chapter involved consideration of the vascular design of the kidneys. On the assumption that the number of capillaries per fundamental kidney unit, or nephron, was constant, it was argued that the number of such units must scale with animal weight to the five-eighths power. This was confirmed using previously published data. Using filtration mechanics, it was also predicted that urine output should scale with animal weight raised to the five-sixths power. This, too, was confirmed with previously published measurements.

Consideration of the design of the lungs in Chapter 5 led to relations between the radius and net length of characteristic capillaries and the net volume and surface area of pulmonary capillaries. From published data on the latter quantities, the radius and net length of characteristic capillaries were determined for various mammals and shown to follow closely the one-twelfth and five-sixths scaling laws predicted by the theory. Consideration of oxygen diffusion further led to the prediction that net capillary area must be proportional to the respiratory membrane area, a result found to be in agreement with measurements. Some interesting aspects associated with transfer of oxygen from the lungs to the pulmonary blood were also discussed in this section.

Certain details of the mechanical design of the heart, not previously considered, were next discussed in Chapter 5. Of particular importance was the prediction that the stress in the walls of the heart is independent of animal size. In engineering terms, this means that the heart of the elephant has the same stress reserve and factor of safety against stress rupture as that of the mouse, even though one pumps more than 6000 times as much blood per minute as the other. The efficiency of the heart was also predicted to be independent of animal size so that the ratio of power produced to power consumed by the hearts of the mouse and elephant are the same.

In Chapter 6, we considered the subject from a different point of view and examined the response characteristics of the cardiovascular system as predicted by a simplified representation of the governing design equation developed in Chapter 3. Our attention was directed specifically toward the human, but the results could be scaled to other mammals using the known similarity of the system. Of particular significance here was the prediction of the variation of cardiac output with increasing heart rate during exercise. The theory indicates that this response is something like that of a simple mechanical system having a mass attached to a parallel-connected elastic spring and viscous element, with a periodically varying force applied to the mass. As the frequency of the periodic force increases, the mass undergoes larger and larger displacements even though the amplitude of the force does not change. When the forcing frequency reaches the natural frequency of the system, a so-called resonant condition exists and the displacements are maximum. Further increase in frequency then results in decreased displacements.

In the case of the cardiovascular system, the present design theory indicates a similar response with increasing exercise and heart rate. However, because reserve capillaries open up and become active during exercise, the increase in heart rate with exercise is accompanied by a decreased viscous resistance, so that the maximum cardiac output at maximum heart rate is indicated to be the result of increased heart rate, resonant dynamic amplification of ventricle displacement, and decreased viscous resistance to blood flow.

When predictions from the theory were compared with published measurements from human subjects, we saw good general agreement with the measured variation of cardiac output with heart rate. Indications from the theory were that the total number of active capillaries increased by a factor of about three during exercise and that cardiac output increased by a factor of about five over the resting value. Interestingly, without the accompanying increase in capillary number, the theory predicts less than a two-fold increase in cardiac output during exercise from the resting value. Another interesting aspect of the design theory is that the increased cardiac output during exercise that results from increased ventricle displacement is predicted to occur without the necessity for any increase in the strength of cardiac muscle contraction. Comparison of the theory with measurement also suggested the possibility that the opening of the reserve capillaries is not a graded response, but rather an all-or-nothing response such that all are open in those parts of the body subjected to exercise states ranging from moderate to strenuous. It was noted, however, that this view would be subject to change if a graded increase in cardiac strength accompanied increasing exercise as a result, for example, of sympathetic nerve stimulation.

7.2 DESIGN METHODOLOGY

Let us now turn from review of the present work to a brief consideration of methodology that we could use if we were involved in an imaginary project of developing the design of the cardiovascular systems for mammals. Assuming we were facing this task for the first time, we may imagine constructing a rough working system and placing it in operation in a proposed mammal of specified weight. By adjusting variables such as heart size, number and dimensions of vascular elements, cardiac cell size, etc., we could, by trial process, presumably obtain some desired level of operation and efficiency of the system. We could then document all details of this final design and prescribe suitable adjustments to its variables so that the system would work in a mammal of any size with the same performance and efficiency.

The prescriptions for change of variables under change of size are, of course, the theoretical scaling laws developed in the present work. In applying these, we may refer to the above experimental system and mammal as the model and any different system and mammal as the prototype. We consider the geometry of the system to involve variables that we can control and, hence, we think of these as independent variables. Other presumably controllable variables such as oxygen-partial pressure in the blood are also considered independent variables. Response quantities such as cardiac output are then determined by the system and can be thought of as dependent variables.

To illustrate the scaling of independent variables, we may consider the length dimensions of the heart which we know scales with animal weight raised to the one-third power. Thus, the relation between any length ℓ_m in the model and the associated length ℓ_p in the prototype system is expressible as

$$\ell_p = \left(\frac{W_p}{W_m}\right)^{1/3} \ell_m$$

where W_p and W_m denote weights of model and prototype mammals. Similar relations can be written for variables defining the vascular geometry. The scaling relation for capillary radius r_c, for example, is

$$r_{cp} = \left(\frac{W_p}{W_m}\right)^{1/12} r_{cm}$$

where subscripts m and p refer to model and prototype, as above.

The scaling of a dependent variable is based on the same procedure, but in this case, the resulting value is that obtained in the prototype rather than specified for it. The cardiac output Q_b, for example, will scale according to the relation

$$Q_{bp} = \left(\frac{W_p}{W_m}\right)^{3/4} Q_{bm}$$

provided the independent variables of the system have all been properly scaled to ensure similarity.

Table 7.1 illustrates typical independent variables involving geometry of large and small vessels and their values such as might be obtained in the above design process by scaling from a model mammal of 2-kg weight to a prototype mammal of 70-kg weight. Also included is the required value for the oxygen pressure in the blood at 75% saturation. Notice that, in the scaling of large vessels, the number of such vessels has not been included as a variable, since our earlier work showed that this must remain constant under change of scale. The large vessels include mainly the connecting arteries from the heart and the large arteries. All of these should scale by the factors indicated for the typical values. The small vessels include the arterioles, capillaries, and venules and their geometry should likewise scale by the same factors as indicated for the typical values.

In the present work, we have not placed any specific requirement on the scaling of intermediate-size vessels, such as the small arteries. The reason is that resistance to blood flow in these vessels does not dominate the net viscous or inertial resistance felt by the heart during pumping. Presumably, a gradual transition from the scaling relations for the large vessels to those for the small vessels would describe the scaling of these vessels.

We see that illustrative predictions for several dependent variables are also given in Table 7.1. The heart rate is included here as a dependent variable on the assumption that its value is fixed by the size of the cardiac cells. As noted above, we regard the value of these variables as predictions

TABLE 7.1.   Values of Various Physiological Variables for a 70-kg Prototype
             Mammal as Scaled from Assumed Values for a 2-Kg Model
             Mammal

Variable	Typical value	
	Model (2 Kg)	Prototype (70 kg)
Independent variables		
Heart dimension (cm)	2.24	7.33
Large vessel radius (cm)	0.32	1.20
Large vessel length (cm)	20.6	50.0
Small vessel radius (mm)	0.0037	0.0050
Small vessel length (mm)	0.477	1.00
Small vessel number	3.3×10^8	3.0×10^9
Cardiac cell dimension (mm)	0.013	0.020
Oxygen pressure at 75% (mm Hg)	54	41
Dependent variables		
Heart rate (beats/min.)	175	72
Cardiac output (ml/min.)	376	5410
Oxygen consumption (ml/min.)	18.8	271
Urine output (ml/min.)	0.046	1.0
Blood pressure in aorta (mm Hg)	120/80	120/80

of the prototype performance from measurements of the model performance.

We have not yet discussed any restrictions on the range of application of the scaling laws developed in the present work and applied here in illustration of a hypothetical design procedure. Restrictions must, of course, apply to the range of animal weights covered because of physical limitations of some kind.

First let us consider the lower limit to body weight that can be expected from the theory itself. We may recall from Chapter 2 that arguments were made for neglecting viscous resistance to blood flow in the larger macroscopic vessels relative to that in the smaller microscopic vessels. These arguments were based on the geometric dimensions and numbers of the two classes of vessels. Since the scaling laws developed in Chapter 3 require the geometry of the macroscopic vessels to vary more under change of scale than the microscopic vessels, we may therefore expect for sufficiently small values of animal weight that the macroscopic vessels will be small enough, relative to the microscopic vessels, to make the argument invalid. To examine this, we may use Eq. (2.26) and the scaling relations (3.29) and (3.30) to show that the ratio of viscous resistance, f_{1v}, in the aorta to viscous resistance in the arterioles, f_{4v}, varies with animal weight raised to the negative one-half power. For the human, this ratio was shown in Table 2.5 to be of the order of 10^{-2}. Using the scaling relation just noted, we may scale this value to other animal sizes and obtain the results shown in Table 7.2.

From the results in this table, it can be seen that the assumption of negligible viscous resistance in the aorta breaks down for animal weights somewhat less than 0.02 kg, that is, for mammals somewhat smaller than the mouse. Thus, since this assumption is a fundamental one in the present design theory, we must limit the theory's applicability to mammals no smaller than, say, 0.01 kg in weight. Interestingly, this limit also appears to be that found in nature. The very smallest adult mammals existing are the small shrews weighing 0.002 kg to 0.003 kg, and these are known to be of special design, with heart rate and oxygen consumption rate differing significantly from scaled values derived from other larger mammals (Schmidt-Nielsen, 1984).

When we next consider the upper limit on the applicability of the scaling laws of the present theory, we find no similar restriction resulting from assumptions of relative importance of flow resistance in the blood vessels. The larger the animal, the better is the assumption of negligible viscous resistance in the macroscale vessels. The better, also, is the assumption of negligible inertial resistance in the microscopic vessels when compared to the macroscale vessels. Upper limits on the size of mammals would therefore appear to arise from other considerations such as locomotion, food supply, or heat dissipation. Indeed, it is unlikely to be simply a matter of coincidence that the largest mammals, the whales, make their home in the sea.

In the above hypothetical design process, we have assumed a single

TABLE 7.2.   Order of Magnitude Estimates of the Ratio of Viscous Resistance in the Aorta f_{1v} to That in the Arterioles f_{4v} for Different Animal Sizes

Animal weight (kg)	Order of f_{1v}/f_{4v}
200	10^{-2}
70	10^{-2}
2	10^{-2}
0.2	10^{-1}
0.02	10^{-1}
0.002	10^{0}

model and imagined varying its design until some desired performance and efficiency were obtained. We then imagined the design of other mammals to be appropriately scaled versions of this one. Alternatively, we may imagine that we start with rough working designs for all mammals and we adjust each to get the desired performance. Since all mammals are constructed of the same material, we would expect that the final design of each would be the same as if it were deliberately scaled from a single model. The scaling laws developed in the present work would likewise apply.

7.3 IMPLICATIONS AND FUTURE WORK

The present work has brought together under one theory important aspects of the design and operation of the cardiovascular system of mammals. As with any physical theory, there are two main requirements to be met if the present theory is to be a successful one. The first is that it consolidate and explain existing measurements and observations in terms of fundamental ideas. The second is that it provide predictions that can guide new experimental measurements and, hence, advance our understanding by either supporting the theory or requiring modification.

Insofar as the first is concerned, we see that a theoretical basis is now provided for those scaling relations of the cardiovascular system that have been established previously from various physiological measurements (Chapters 3, 4, and 5). Respiration measurements on tissues from organs of the mouse and dog have also been brought within the framework of the theory (Chapter 5), as have response measurements of the cardiovascular system during exercise (Chapter 6).

Associated with the theoretical basis for these measurements is an increased understanding of what they mean and why they follow the pattern that they do. For an example, we may recall from Chapter 5 that the number of fundamental kidney units, the nephrons, varies with animal weight raised to the 0.62 power, while the net urinary flow varies with animal weight raised to the 0.82 power. Thus, the output of a single nephron must vary with animal weight raised to the 0.2 power and, hence, as first noted by Adolph (1949) without explanation, must increase with animal size. But why the difference in nephron output? The question is now answered by the theory (Chapter 5). The number of capillaries in a nephron is assumed in the present theory to be the same for all mammals, but their length is predicted to vary with animal weight raised to a power five twenty-fourths, or approximately 0.2. Since the urinary flow per nephron is proportional to the product of the number of its capillaries and their length, the observed difference in nephron output is thus predicted and explained in terms of the difference in characteristic capillary length among mammals.

The theory also provides us with other equally interesting explanations of observed characteristics. Thus, for instance, the decrease in the resting heart rate of mammals accompanying increase in size is an observation predicted by the theory as a result of increasing cardiac cell size, heart rate being inversely proportional to the square of cardiac cell size. The increase in oxygen-consumption rate accompanying increase in animal size is likewise an observation predicted by the theory as a result of increase in both the number and size of average body cells, the rate being as the product of the two.

With respect to the second requirement for a successful theory, we may note that new scaling relations have been established here that require experimental examination. For example, the linear dimensions of the cardiac cells and their number are predicted by the theory to vary with animal weight raised to the one-eighth and five-eighths powers, respectively (Chapter 3). Such variations should be measurable and, if so done, would provide important new findings. Linear dimensions and number of cells of other organs are also predicted and these should also be capable of experimental investigation (Chapter 5). Of particular interest are the brain cells whose linear dimensions are predicted to be the same for all mammals.

It has also been assumed in the development of the theory that the rate of propagation of the contraction signal in cardiac fiber varies with fiber diameter raised to the two-thirds power (Chapter 3). This condition could presumably be examined experimentally in some detail. The results would be of significance both to the present theory as well as to understanding the fundamental processes involved in the signal propagation.

One of the particularly interesting predictions of the present theory is the non-uniform variation of the geometry of the vascular system associated with mammals of varying size (Chapter 3). Indeed, were it not for this non-uniform variation, no similarity in response of the cardiovascular system would exist and different-size mammals would operate at different degrees of efficiency. We have already seen some reasonably direct evidence of this variation from studies of the aortic sectional area (Chapter 4) and from studies of pulmonary capillaries (Chapter 5). A productive line of further experimental work could involve additional studies of both systemic and pulmonary vessels, using the scaling predictions of the present theory as a guide.

In Chapter 6, we considered a solution to the basic design equation governing ventricular displacements under certain simplifying assumptions. This solution provided us with some interesting insight into the working of the cardiovascular system during exercise. A further study of solutions to this equation under less restrictive assumptions could perhaps provide additional understanding of other details of the system's operation and allow a more accurate representation of cardiac response over a heart cycle.

Taken altogether, the present theory and description can be considered to broaden our view of the design and operation of the cardiovascular system, to bring some additional order and explanation to the various measurements made over the years in attempts to understand it, and to provide guidance to us for new investigations that could lead to increased understanding. In our quieter moments, it can also serve to remind us of the truly marvelous organization and operation of the system in all mammals.

REFERENCES

ADOLPH, E. F. 1949. Quantitative relations in the physiological constitutions of mammals. *Science* 109: 579–85.

SCHMIDT-NIELSEN, K. 1984. *Scaling: Why is Animal Size So Important?* Cambridge: Cambridge University Press.

A

Logarithms and Scaling Equations

Logarithms are well known in algebra and are simply the exponents of a base number that are required to give specified values. For example, if Y and Z denote specified numbers, we may express these using a base number 10 as

$$Y = 10^a, \quad Z = 10^b$$

where a and b denote appropriately selected exponents. The logarithms are then written as

$$\log Y = a, \quad \log Z = b$$

From standard manipulation of exponents, we find that Y raised to the n-th power is expressible by the relation

$$\log Y^n = \log 10^{na} = na = n \log Y$$

that is, the logarithm of Y raised to the n-th power is equal to n times the logarithm of Y. Similarly, for the product YZ, we have

$$\log YZ = \log (10^{a+b}) = a + b = \log Y + \log Z$$

so that the logarithm of the product is equal to the sum of the logarithms of the individual terms.

Now let us consider a power law, or scaling equation, of the form

$$Y = K_c W^n$$

where Y is some dependent variable of interest, W is the independent variable and K_c is a coefficient. From the above considerations, we may write the power-law relation as

$$\log Y = \log K_c + n \log W$$

This equation is in the form of that for a straight line. When values of log Y are plotted against the associated values of log W using linear-scale axes, the resulting straight line will have a slope equal to n. The value of log K_c will also equal the value of log Y at log $W = 0$. Alternatively, we may use logarithmic-scale axes and plot values of Y against the values of W directly to get the straight line with slope equal to n and the value of K_c equal to Y for $W = 1$.

As an example, we may consider the simple scaling equation

$$Y = 4 W^2$$

Selected values of W, Y, log W and log Y are tabulated below in Table A.1.

These values have been plotted on the graph in Fig. A.1. The upper horizontal scale of this graph is a linear scale defining values of log W that

TABLE A.1

W	Y	$\log W$	$\log Y$
4	64	0.602	1.806
12	576	1.760	2.760
50	10,000	1.699	4.000

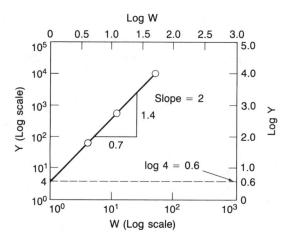

Fig. A.1 Graph of quadratic power law $Y = 4W^2$ illustrating use of logarithms and logarithmic-scale axes

have been plotted, and the right vertical scale is a linear scale defining values of log Y that have been plotted. Using these axes, the resulting straight line is seen to have a rise of 1.4 over a distance of 0.7, giving a slope equal to 2. At the value of log $W = 0$, the value of log Y is also seen to equal 0.6. This corresponds to a value of Y of 4. Thus, we see that the slope of the line is the exponent 2 in the power-law relation, and the intercept value at log $W = 0$ is the logarithm of the coefficient of the power law.

In this same graph, the lower horizontal axis has been labeled with the values of W associated with the log W values on the upper scale. The lower scale is thus a logarithmic scale. Similarly, the left vertical scale has been labeled with values of Y on the right vertical scale, so that this also is a logarithmic scale. A value of, say, 5 on this scale does not fall midway between the 10^0 and 10^1 segment, but rather at about 70% (log 5 = 0.7) of the distance from 10^0 to 10^1.

B

Scaling Laws for Volumes and Areas

We consider here scaling laws for volumes and areas of geometrically similar bodies, that is, bodies of the same shape but not necessarily the same size. For simplicity, we first consider the parallelepiped shown in Fig. B.1, having length a_1, width a_2 and height a_3. The volume B of this solid is determined by the product $a_1 a_2 a_3$ which may be written

$$B = a_1{}^3 \left(\frac{a_2}{a_1} \frac{a_3}{a_1} \right) \tag{B.1}$$

Now, for similarly shaped objects, the ratios a_2/a_1 and a_3/a_1 will be the same, so that this relation may be expressed as

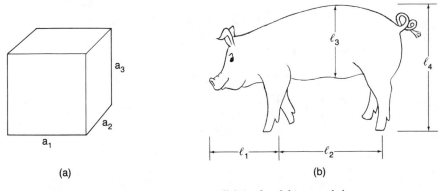

(a)   (b)

Fig. B.1 Linear dimensions of (a) parallelpiped and (b) general shape

$$B \propto a_1{}^3 \tag{B.2}$$

where α denotes proportionality under size change. Thus, if the side a_1 is doubled, the volume will increase by a factor of eight. Of course, doubling side a_1 also means that sides a_2 and a_3 will double, since the shape is required to remain the same.

For objects made of the same material, their weight W will be proportional to their volume. Thus, replacing B with W in the above relation and taking the cube root, we have

$$a_1 \propto W^{1/3} \tag{B.3}$$

The length a_1 therefore varies with the weight raised to the one-third power. Since lengths a_2 and a_3 must also vary in this way, we can think of the length a_1 as a characteristic length representative of all lengths (or dimensions) of the body.

Next consider the surface area S_a of the parallelepiped. This is expressible as $2\,a_1 a_2 + 2\,a_1 a_3 + 2\,a_2 a_3$, or as

$$S_a = a_1{}^2 \left(2\frac{a_2}{a_1} + 2\frac{a_3}{a_1} + 2\frac{a_2}{a_1}\frac{a_3}{a_1}\right) \tag{B.4}$$

For similar shapes, the terms in parentheses are constant. Hence, using Eq. (B.3), we have

$$S_a \propto a_1{}^2 \propto W^{2/3} \tag{B.5}$$

that is, surface area varies as the square of the characteristic length, or, equivalently, as weight to the two-thirds power.

The above results apply also to objects of general shape made of the same material. Thus, for the general shape shown in Fig. B.1, the various length dimensions are ℓ_1, ℓ_2, etc. For similar shapes, the ratios ℓ_2/ℓ_1, ℓ_3/ℓ_1, etc., must be the same from one body to another. Hence, we have the volume, weight, and surface area expressible in terms of, say, the characteristic length ℓ_1 as

$$B \propto \ell_1{}^3, \ W \propto B, \ S_A \propto \ell_1{}^2$$

from which we have the scaling results

$$\ell_1 \propto W^{1/3}, \ S_A \propto W^{2/3}$$

analogous to relations (B.3) and (B.5).

C

Dimensional Analysis and Physical Scaling Laws

Dimensional analysis involves the study of the physical dimensions, or units, of variables in equations and is useful in deriving scaling laws for systems. It is based on the fundamental concept that any relation between physical variables must be expressible in a manner independent of the particular units used to measure the variables. Any such relation must therefore be expressible in terms of independent dimensionless ratios of the variables. Moreover, by requiring these ratios to remain constant while varying the size of the system, scaling laws for the variables can be investigated.

If L denotes any unit of length (cm, ft, etc.) and T any unit of time (sec, min., etc.), we see that the units of, say, velocity may be expressed as L/T and those of acceleration as L/T^2. Similarly, if F denotes any unit of force (dyne, pound, etc.), the units of, say, pressure may be expressed as F/L^2.

In mechanical systems, we are usually concerned with units of length, time, force, and mass. However, since force is equal to the product of mass and acceleration by Newton's Second Law of Motion, we see that the units of force and mass are not independent, and we have the relations

$$F = ML/T^2 \text{ and } M = FT^2/L$$

Thus, either mass or force may be considered the fundamental unit.

C.1   EXAMPLE

For a simple example, we may consider the relation connecting the bearing pressure exerted on the leg of an animal to the geometric and physical variables on which it depends. The problem is illustrated in Fig. C.1. We have already identified the quantity of interest as the bearing pressure, and we denote it by P. This pressure must depend on the geometric dimensions, say, ℓ_1, ℓ_2, ℓ_3, etc., describing the size and shape of the animal and on the average weight per unit volume w of the material (tissue, bone, etc.) comprising it. Now the pressure has units of F/L^2, the geometric dimensions have units of L, and the weight parameter has units of F/L^3. Thus, on forming independent dimensionless ratios, we find

$$P/w\,\ell_1,\ \ell_2/\ell_1,\ \ell_3/\ell_1,\ \text{etc.}$$

The equation connecting these dimensionless variables may be written in general terms as

$$P/w\,\ell_1\ =\ f(\ell_2/\ell_1,\ \ell_3/\ell_1)$$

where $f(-)$ denotes a function of the indicated variables, and we have, for simplicity, only included three geometric dimensions.

We now consider a change in the size of the animal, with all lengths ℓ_1, ℓ_2, ℓ_3 changing by the same amount. In this case, the ratios ℓ_2/ℓ_1, ℓ_3/ℓ_1 will remain unchanged so that the size of the animal will change but not its shape. Because of the constant values of the length ratios, we see from the

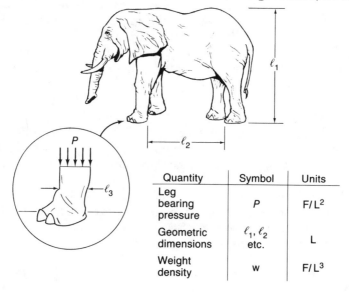

Quantity	Symbol	Units
Leg bearing pressure	P	F/L^2
Geometric dimensions	ℓ_1, ℓ_2 etc.	L
Weight density	w	F/L^3

Fig. C.1   Illustration of leg-bearing pressure and associated variables above equations that the ratio $P/w\ell_1$ must also remain constant. In addition, with identical construction materials used for animals of different size, the weight parameter will be the same, in which case we see that the ratio P/ℓ_1 must remain constant. We therefore see that bearing pressure will vary (or scale) directly with change in the characteristic length ℓ_1. If we use the result from Appendix B that characteristic lengths are proportional to body weight to the one-third power, we see that bearing pressure must satisfy the scaling relation

$$P \; \alpha \; W^{1/3}$$

The scaling law that we establish is that leg bearing pressure will, for similarly shaped animals made of identical construction materials, vary directly with the cube root of the body weight.

D

Some Physical
Concepts

D.1 PIPE FLOW

The movement of fluid (liquid) in pipes, or cylindrical vessels, is characterized broadly by the flow Q, or ratio of volume of fluid moved to the time
taken. The units of flow are length and time, in the form L^3/T. Associated
with the flow is the average velocity V of the fluid motion. This is defined
as the ratio of fluid flow to cross-sectional area of the vessel. Its units are
length and time, in the ratio L/T. The average acceleration $\dot{V}$ is the rate of
change of the average velocity, defined as the ratio of change in mean velocity to time taken with units of length and time in the ratio L/T^2. The
rate of change of the flow Q is the product of the average acceleration and
the cross-sectional area of the vessel. The units are length and time in the
form L^3/T^2.

These quantities are illustrated in Fig. D.1 where an imaginary disc
of fluid is assumed to move through the vessel with average velocity at
each instant. At a time equal to 2 sec, the flow in the tube is equal to 18
cm³/sec and the average velocity is determined as 10.2 cm/sec. At a later
time of 2.5 sec, the flow in the tube is equal to 20 cm³/sec and the average
velocity is determined as 11.4 cm/sec. An approximate calculation for the
average acceleration is then seen to give the value of 2.4 cm/sec². The calculation is approximate because the time interval and associated change
in velocity should be very small for an estimate of the average acceleration at, say, the time of 2 sec. The rate of change of the flow $\dot{Q}$ is calculated

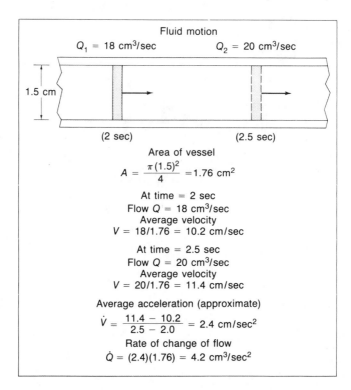

Fig. D.1 Illustration of calculations of fluid motion

from the average acceleration as 4.2 cm³/sec². Although not shown in the calculations in Fig. D.1, we may also estimate the distance that the imaginary disc travels in the time interval of 0.5 sec. A representative velocity over the time interval is the simple mean given by (10 + 11.5)/2, or 10.8 cm/sec. The distance traveled is then 10.8.(0.5) or 5.4 cm.

The driving force for fluid flow in a vessel is the net end pressure, defined as the difference between the axial force per unit of cross-sectional area acting at the entrance and exit ends of the vessel. More generally, fluid pressure is defined as the force per unit of area acting perpendicular to any real or imagined fluid surface. The units of pressure are force and length in the ratio F/L^2.

The viscosity of a fluid refers to its internal resistance to relative slipping of layers of the fluid. The viscosity coefficient μ measures this resistance by relating the tangential force per unit area associated with slipping of two layers of fluid to the ratio of relative velocity to perpendicular distance between the layers. The viscosity coefficient thus has units of FT/L^2. In the case of fluid flow through a cylindrical vessel of length L and radius r, the viscosity coefficient is the material parameter involved in relating the net driving pressure f_r for this resistance and the average fluid velocity V through the tube; that is

$$f_r = 8\mu \frac{L}{r^2} V \tag{D.1}$$

This equation is a version of the famous Poiseuille equation.

The inertia of a fluid refers to its resistance to acceleration. The material parameter involved is the mass density ϱ of the fluid, defined as the amount of mass present in a unit volume of the substance. The driving pressure f_m needed to accelerate fluid in a vessel of length is given by

$$f_m = \varrho L \dot{V} \tag{D.2}$$

where $\dot{V}$ denotes the average acceleration.

These concepts are illustrated in Fig. D.2 for the case of a fluid (blood) flowing through a vessel of 0.20 cm diameter and 100 cm length. The average velocity and acceleration are 2 cm/sec and 10 cm/sec². The driving pressure is calculated to be 5850 dynes/cm², which is equivalent to the pressure at the base of a column of mercury of height 4.39 mm. Thus,

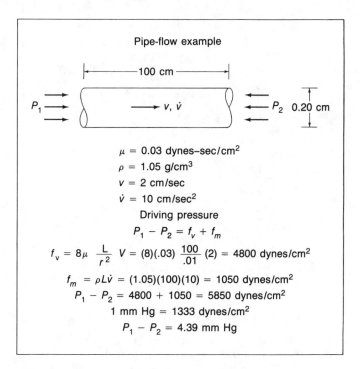

Fig. D.2   Illustration of calculations for blood flow in a small-diameter vessel

the pressure is also expressible as 4.39 mm Hg. The pressure unit of mm Hg is sometimes written as Torr, that is, 1 mm Hg = 1 Torr.

D.2 ELASTICITY

The elasticity of a body refers to its springiness. It can be measured by the elastic modulus E, which is a material parameter relating the tensile force per unit of cross-sectional area of a bar or fiber to the associated change in length per unit of length (see Fig. D.3). If, for example, the relation is expressible as a power law with exponent b, we have

$$\frac{F}{A_o} = E\left(\frac{\Delta L_o}{L_o}\right)^b \tag{D.3}$$

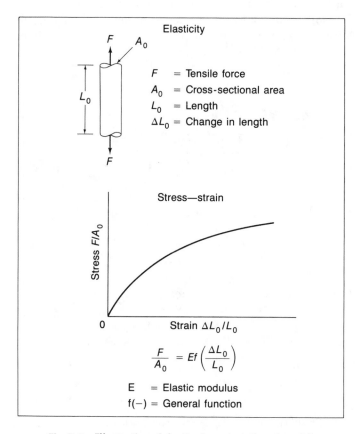

Fig. D.3 Illustration of elastic characteristics of a solid where F denotes the tensile force, A_o the cross-sectional area, ΔL_o the change in length and L_o the initial length. The units of E are F/L^2.

D.3   DIFFUSION

Diffusion refers to the net movement of material as a result of a difference in concentration. In the case of a permeable membrane separating two concentrations, the rate of movement of material J across the membrane depends on the membrane area A, its thickness h, and the concentration difference according to the equation

$$J = D_s \, \frac{A_m}{h_m} \, (C_1 - C_2) \tag{D.4}$$

where C_1 and C_2 denote concentrations (mass per unit volume) of the substance on opposite sides of the membrance and D_s denotes the diffusion constant. This equation expresses a version of Fick's law of diffusion. Since J has units of M/T, A_m has units of L^2, h has units of L, and $C_1 - C_2$

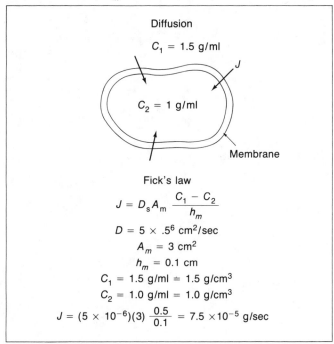

Fig. D.4   Illustrative calculation of diffusion of sugar across a membrane in an aqueous solution

has units of M/L^3, it can be seen that the diffusion constant must have units of L^2/T.

In the case of a gas dissolved in a liquid, the concentration C, expressed as volume of gas per volume of solution, is proportional to the partial pressure P_g (external pressure needed to keep the gas from escaping). Fick's law is then expressible as

$$ J = D_G \frac{A_m}{h_m} \Delta P_g \qquad\qquad (D.5) $$

where J denotes the rate of diffusion, expressed as volume of gas per unit time, ΔP_g denotes the difference in partial pressures on opposite sides of the membrane, and D_G denotes the diffusion constant for the gas.

Fick's law is illustrated in Fig. D.4 in the calculation of the rate of diffusion of sugar across a membrane in an aqueous solution. For inside and outside concentrations of 2 and 2.5 grams of sugar per milliliter of solution, respectively, the diffusion rate is seen to be 7.5×10^{-5} g/sec, which is equivalent to 6.48 g/day.

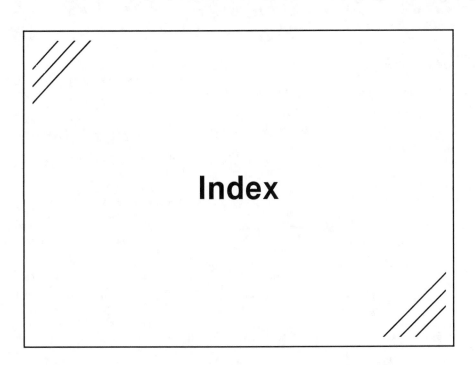

Index

Adolph, E. F., 103, 116, 118–120, 132, 158, 160
Altman, P. L. 88, 91, 100
Amman, A., 133
action potential, 36
allometric equation, 3
alveoli, 121
alveolar ducts, 121–22
aorta:
  flow resistance, 62
  pressure in, 21, 138–39
  radius, 79
arteries, 20–21
arterioles, 50–52, 55, 115
atrial contraction, 63
atrium, 20, 34, 136
A-V node, 37
average body cell, 69–70, 74
average heart cell, 69
average organ cell, 103

Benedict, F. G., 10, 18
blood:
  composition, 22
  density, 14, 32
  functions, 21
  hematocrit, 23
  viscosity, 14, 32
  weight, 3–4

blood acceleration, 84
blood cells, 23
blood circulation, 19, 86–87
blood flow:
  average, 7, 21, 81–82, 140
  bronchial, 21, 24, 50, 121
  organs, 110
  pulmonary, 121
  resistance, 14, 54, 136
blood motions, 81–86
blood pressure, 84–86, 147
blood velocity, 84–86
Blount, S. G., Jr., 148
Bock, A. V., 140, 148
body surface area, 9–13, 17
Bonner, J. T., 17–18
Borrero, L. M., 59, 133
Bowman's capsule, 115
brain, 108–110
breathing rate, 121
Brody, S., 4, 9, 10–11, 18, 103, 108–110, 133
bronchial flow, 21, 24, 50, 121
bronchioles, 121
Brown, F. A., Jr., 87, 101
Bullard, R. W., 37, 60

Campbell, K. B., 73–74, 85, 101
capillary:
  characteristic radius, 123
  density, 98–100
  net length, 123–24
  scaling law, 71, 124
  net volume, 125
  net lateral area, 125
capillaries:
  pulmonary, 20, 24
  systemic, 19, 24, 52
  glomerular, 115
  peritubular, 115
carbon dioxide, 28
cardiac cells, 34, 69–70
cardiac cycle, 35, 136
cardiac excitation, 36
cardiac muscle mechanics, 40–48
cardiac output, 7, 13, 21, 81–82, 140
cardiac response:
  resting, 137
  exercise, 140
cardiovascular system:
  essential features, 1, 19–21
  mathematical models, 61

cardiovascular sys-
   tem (cont.)
  similarity in design, 2,
     8
cells, 23, 40, 69–70, 74,
   103
circulation of blood, 19
circulation time, 86–87
Clark, A. J., 7, 18, 76, 79,
   90, 100
Cokelet, G. R., 33, 59
contractile element, 40
contractile factor, 41–48
contractile force, 41–48,
   135–36

Daugherty, R. L., 55–56,
   59
Davson, H., 113, 133,
   140, 148
dead-space volume, 127
diastole, 35
diffusion, 175
Dill, D. B., 148
dimensional analysis, 13,
   167
dissolved products,
   30–31
Dittmer, D. S., 88–89, 91,
   100
Downie, E. D., 108, 133
Downing, S. E., 44–45,
   59

Eckstein, R. W., 100
Eggleton, M. G., 113,
   133, 140, 148
elastic element, 40
elastic modulus, 14, 48,
   64, 174
elasticity of a body, 174
engineering description,
   24, 41, 44
engineering design
   model, 62

Fahraeus-Lindquist ef-
   fect, 33
fibrils, 34
Fick's Law, 30, 70, 175
Filley, G., 148
Fineburg, M. H., 100
Folling, A., 148
Franzini, J. B., 55–56, 59
Friedman, J. J., 52, 59
Fuhrman, F. A., 104–5,
   133

Gehr, P. D., 123–26, 129,
   133
geometric scaling rela-
   tions:
  aortic radius, 79
  end-diastolic volume,
     77
  general, 68, 71
  length of left ventricle,
     76
glomerulus, 114–15
governing equation, 66,
   136
Gregg, D. E., 85, 100
Grover, R. F., 148
Guerra, E., 92, 100
Gunther, B., 92, 100
Guyton, A. C., 31, 59, 69,
   74, 95, 100, 128,
   133, 145, 148

Hamilton, W. F., 85, 101
Harvey, William, 19
Hill, A. V., 46, 59
Holt, J. P., 77, 100
hoop-force equation, 63
Hoppeler, H., 133
heart:
  cellular structure of,
     34, 69
  force, 46
  left side, 19, 63
  mechanical design of,
     130
  rate, 4, 15, 40, 89, 140
  right side, 19, 63
  sectional view, 34–35
  weight, 2, 4, 7, 67, 69
  work and efficiency of,
     132
heat loss, 9, 12
heat production, 6, 7, 9,
   17
heat theory, 9–13
hematocrit, 23
hemoglobin, 23
Hurxthal, L. M., 148
Huxley, J. S., 3, 18

inertia of fluid, 173
isometric contraction, 35,
   47, 66
isometric phase, 40
isometric response, 46
isometric stage, 42

isotonic phase, 41
isotonic stage, 42

Jack, J. J. B., 70, 74

Kenner, T., 16, 18, 85,
   100
kidneys, vascular design
   of, 114
Kines, H., 77, 100
Kleiber, M., 7, 10, 18
Krogh, A., 98, 100
Kunkel, P. A., 116, 133

Lambert-Teiser Scaling,
   9, 13, 16
Larimer, J. L., 95–97, 101
leg bearing pressure, 169
Li, J. K. J., 73–74, 85,
   101
Lloyd, T. C., Jr., 27, 59
logarithms, 161
lung design, 121
lung volume, 126–27

macroscale vessels, 57
Maloiy, G. M. O., 133
Martin, A. W., 104–5,
   133
mass density, 32
Mathieu, O., 133
McDonald, D. A., 87, 100
McMahon, T., 16–18
mechanical design of
   heart, 130
membrane:
  pulmonary, 128
  respiratory, 128
mesenteric system, 111
microcirculation, 52
microscale vessels, 57
modeling theory, 13
Munro, H. N., 108, 133
muscle shortening, 45
Mwangi, D. K., 133

nephra number, 115
nephra scaling relation,
   116
nephron, 114, 120
Noble, D., 70, 74
Noordergraaf, A., 19, 52,
   56, 59, 61, 73–74,
   85, 101

oxygen:
  consumption rate, 4–5,
     8, 17, 70–73, 140

degree of saturation of,
    26, 140
exchange, 23
measurements, 104
partial pressure of, 23,
    26
transfer, 24, 126
transport, 23
utilization, 8
oxygen-consumption
    rate:
during growth, 88–89
measurements, 4
scaling law, 5–6, 71–72
oxygen-dissociation
    curve, 26–27
oxygen-partial pressure:
blood, 94–98
cells, 93
oxygen uptake of organs,
    103
oxyhemoglobin, 23

Pappenheimer, J. R., 31,
    59, 117, 133
parasympathetic nerves,
    38
Pennycuik, P., 98–99,
    101
pipe flow, 171
platelets, 23
Poiseuille equation, 33,
    173
portal vein, 111
precapillary sphincters,
    51–52
pressure equation, 65
Prosser, C. L., 87, 101
Prothero, J. W., 108,
    112–13, 133
pulmonary flow, 121
pulmonary membrane,
    48, 128
pulmonary vascular bed,
    19, 21, 58

Reeves, J. T., 144–45,
    148
Remmers, J. E., 126, 129,
    133
Renkin, E. M., 59, 133
respiratory membrane,
    128
Rhode, E. A., 77, 100
rhythmic electrical dis-
    charges, 38

S-A node, 36–37
Sagawa, K., 43, 46, 59,
    61, 74, 132–33,
    144, 148
Sarrus-Rameaux Theory,
    9, 11–12, 16
scaling equations, 161
scaling of intermediate-
    size vessels, 155
scaling laws:
areas, 165
average body cells, 71
blood motions and
    pressures, 83–86
blood weight, 3
capillary density, 98
cardiac cells, 71
cardiac output, 7, 82
circulation time, 86
during growth and ag-
    ing, 88–92
heart rate, 4, 71
heart weight, 3
investigations of,
    13–16, 167–169
lung volume, 127
nephra number, 116
organ blood flow, 111
organ cells, 107
oxygen-consumption
    rate, 5, 72, 104
oxygen partial pres-
    sure, 93–98
restrictions on, 156
urine flow, 119
vascular system, 71,
    79, 123–26
ventricular stress, 131
ventricular power, 132
volumes, 165
scaling relations, 70, 72,
    106–7
Schmidt-Nielsen, K., 11,
    18, 27, 59, 95–99,
    101, 126, 133, 157,
    160
Selkurt, E. E., 37, 60
similarity requirements,
    66
Smith, R. E., 108, 133
Sonnenblick, E. H.,
    44–47, 59–60
Stahl, W. R., 121, 133
Starling, Ernest, 41

Starling Law of the
    Heart, 41
stratification, 130
sympathetic nerves, 38
systemic blood flow, 52
systemic vascular bed,
    19, 21, 48, 58
systemic vascular
    system:
idealized description
    for, 50
resistance to blood
    flow, 54–57
systole, 35

Taylor, C. R., 133
Tenney, S. M., 126, 129,
    133
tidal volume, 126–27
trachea, 121
Tsien, R. W., 70, 74

urine flow, 117–20

valves, 34
Vancaularet, C., 148
vasa recta, 115
vascular beds, 48–49
vascular design of kid-
    neys, 114
veins, 19, 21
vena cava, 54
venous saturation, 140
ventricle, 34, 48, 61, 63,
    136
ventricular:
contractions, 138
force, 136
outflow, 66, 135
pressure, 35, 138, 140,
    146
response characteris-
    tics, 138
stress, 131
volume, 135
ventricular equation, 63
venules, 50
viscosity, 32, 172
viscosity coefficient, 32,
    172

water, exchange of,
    30–31
Weibel, E. R., 130, 133
windpipe, 121
Woodbury, R. A., 85, 101